MICROWORLDS

ALSO BY STANISLAW LEM

STANISLAW LEM
MICROWORLDS

WRITINGS ON SCIENCE FICTION AND FANTASY

Edited by Franz Rottensteiner

A Helen and Kurt Wolff Book
A Harvest/HBJ Book
Harcourt Brace Jovanovich, Publishers
San Diego   New York   London

Requests for permission to make copies of any part of the work
should be mailed to: Permissions, Harcourt Brace Jovanovich,
Publishers,
Orlando, Florida 32887.

Library of Congress Cataloging in Publication Data
Lem, Stanisław.
Microworlds: writings on science fiction and fantasy.
"A Helen and Kurt Wolff book."
Bibliography: p. 279
I. Science fiction—History and criticism—Addresses,
essays, lectures. 2. Fantastic fiction—History and
criticism—Addresses, essays, lectures. I. Rottensteiner,
Franz. II. Title.
PN3433.8.L4   1984   809.3'876   84-12837
ISBN 0-15-159480-5
ISBN 0-15-659443-9 (Harvest/HBJ : pbk.)

Designed by Mark Likgalter
Printed in the United States of America
First Harvest/HBJ edition 1986

A B C D E F G H I J

CONTENTS

INTRODUCTION

It was toward the end of the 1960s that I began corresponding with Stanislaw Lem. I had been a voracious reader of science fiction for many years, although I disliked most of what I read and saw it as a waste of the form's potential. Perhaps it was this similarity in our views that Lem found attractive.

Science fiction differs from other popular genres in that its readers are frequently articulate, eager to meet and talk with other science-fiction fans. There is a whole science-fiction subculture, with hundreds of amateur magazines, or "fanzines," devoted to science fiction, its authors, and its audience. These magazines, most with a circulation of only a few hundred copies, are found not only in the United States, where they started, but all over the world, even in Communist countries. Since the early 1960s I myself have edited such a fanzine, called *Quarber Merkur;* it is devoted to the analysis of science fiction and fantasy writing and is rather critical of them. At that time I knew of Lem, but I considered him only one science-fiction writer among many, though perhaps the most important in Eastern Europe; I had read very little of his work. In Germany he was little more than a name; few of his books had appeared in German, mainly in East Germany. His first science-fiction novel, *Astronauci* (1951; The Astronauts), had been widely translated, and a few other works had appeared in France and Italy, mostly in atrocious translations. That was all.

In 1968 I published a review of an East German translation of Lem's novel *The Invincible* in my magazine and sent the author a copy, without comment. In response, Lem wrote me a long and extremely interesting letter in German. That was the beginning of a long correspondence; by now Lem's letters to me fill three large files. They constitute the most detailed documentation in the West of Lem's thoughts, activities, and international career since 1968. From his letters I recognized a truly remarkable mind, and when I became a science-fiction editor in West Germany in 1970 I was able to publish him. Then it occurred to me that I might do more for Lem if I became his literary agent.

Early in our correspondence, Lem indicated that he was planning to write a study of science fiction but was having difficulty obtaining source materials. I sent him what I considered interesting and drew his attention to a number of writers, among them Cordwainer Smith, Philip K. Dick, J. G. Ballard, C. M. Kornbluth, and Philip José Farmer. Aside from supplying some works of science fiction and also some of the few then existing books about science fiction (especially the criticism of Damon Knight and James Blish), I made no attempt to influence the shape of Lem's book, nor would any such endeavor have been successful with a writer like Lem. (Curiously, some science-fiction writers later implied I had unduly influenced Lem or even made him up.)

The result of Lem's efforts was finally published in 1971 as *Fantastyka i futurologia* (Science Fiction and Futurology). It is both a rigorous investigation of the theoretical basis of science fiction and a detailed analysis of many of its major topics and literary themes. The first volume in particular contains some highly theoretical reasoning that is without precedent in other books on science fiction, most of which are historical, biographical, or bibliographical in character. So far Lem's book has appeared outside Poland only in German and (in abridged form) in Hungarian. Two chapters have been published in English in the journal *Science-Fiction Studies*, "The Time-Travel Story and Related Matters of Science-Fiction Structuring" and "Metafantasia: The Possibilities of Science Fiction." Both give an indication of the freshness and originality of Lem's approach and also shed light on his own science fiction.

While Lem was writing *Fantastyka i futurologia*, we corresponded a great deal, and in his letters Lem provided extensive explanations of what he was doing. Later I published some of these letters as separate articles. "On the Structural Analysis of Science Fiction" had its genesis in a lengthy letter; it is the most succinct statement of the aims of Lem's book. Lem also wrote many reviews and essays for my magazine, and I translated many of Lem's writings for Australian publications like John Foyster's *Journal of Omphalistic Epistemology* and especially Bruce Gillespie's *Science Fiction Commentary.* These writings proved quite controversial for science-fiction buffs, especially the long essay "Science Fiction: A Hopeless Case—with Exceptions," a more polemical version of a chapter from *Fantastyka i futurologia.*

Lem has an insatiable thirst for knowledge and more of a philosophical than a poetic bent; scientific and philosophical inquiry has always played an important part in his work. Even in his fiction there is a strong essayistic element. Learned disquisitions are frequently woven into the plot, and if anything this practice has grown stronger with the passage of time. The stories in the various cycles (such as the Ijon Tichy tales, the Pirx stories, and the philosophical tales of the *Cyberiad*) become more complex with time; sometimes they carry so heavy an intellectual load that the story is in danger of being smothered. Moreover, Lem leans increasingly toward forms that are hybrids of fiction and nonfiction. *His Master's Voice,* a novel of science, is actually a brilliant essay on the limits of human knowledge, the process of cognition, and the moral responsibility of the scientist. It was followed by fictions that do away altogether with conventional characters and narrative. *A Perfect Vacuum* is a collection of reviews of nonexistent books; *Imaginary Magnitude* brings together introductions to equally nonexistent works.

So it is hardly surprising that Lem should have made a critical study of the problem that interests him most, that of the scientific and literary foundations of his own and others' writings. Given the vagaries of translation, however, very little of Lem's criticism is available in English, and most of what is available deals with science fiction, the genre which Lem himself favors most of the time. Of course, the practice of science fiction is an important subject of Lem's nonfiction writing, but it is only one of many. Lem's interests range from cybernetics and artificial intelligence to cosmology and cosmogony, genetic engineering, the creation of simulated environments, literary theory and the reception of literary works, and indeed everything pertaining to the future of man and his civilization.

In *Dialogi* (1957; Dialogues) Lem discussed, in the form of Socratic dialogues between a Berkeleyan Hylas and Filonous, the amazing prospects of the young science of cybernetics. His *Summa Technologiae* (1964), perhaps his most important discursive work, is a futurological treatise unlike anything else on the subject. Instead of presenting the usual catalogue of wonderful or horrible things that the future has in store, Lem selects certain ideas to pursue to their outermost limits—the problem of cosmic civilizations, the evolution of artificial intelligence, the genetic remodeling of man, the creation of worlds, stellar engineering; or he formulates daring hypotheses about the breeding of information or the total reconstruction of reality.

In *Filozofia przypadku* (1968; The Philosophy of Chance), Lem turned to quite another question: why are works of literature received differently in different ages and different cultures, being highly esteemed at certain times and held in low regard at others? Here Lem tried to arrive at an empirical theory of literature that would take into account such temporal and cultural factors. The book also contains a spirited polemic against structuralism, a polemic that is continued in *Fantastyka i futurologia,* in which Lem applies to science fiction the theories elaborated in the earlier, more general volume.

Lem's relationship with science fiction is a love-hate relationship. Although much of his writing can only be called science fiction, for a long time Lem was not familiar with what other science-fiction writers were doing (he has often declared that he lived in Poland like Robinson Crusoe on his island). Verne and Wells he read in his youth, but modern Western science fiction was unknown to him until the late 1950s, when he read some of it, mostly in French translation, following his publication in France by Denoël. Only much later was he able to read more widely, and the more science fiction he read, the more he was disappointed. This disappointment is reflected in many of his autobiographical pieces and, of course, in his criticism. In the late sixties, Lem decided to put his ideas about science fiction into systematic form. The result was *Fantastyka i futurologia,* a major study of Western science fiction.

For a time Lem played the part of a missionary in the science-fiction world, but today he feels he was wasting his time trying to reform science fiction by criticism, and he has virtually stopped writing about it (and reading it), although he still contributes occasionally to publications like *Science-Fiction Studies.* To judge from the reaction to

Lem's essays, few understood him. But some people at least seem to have understood him: in 1976 the Science Fiction Writers of America revoked Lem's honorary membership, following publication by an American press service of excerpts from an article of his in a German newspaper, sharply critical of science fiction. Officially, Lem's membership was withdrawn on technical grounds. Had Lem been less critical of science fiction, of course, the SFWA officers would have had no reason to read the bylaws of their own organization, and Lem would not have been treated so shabbily. (The whole affair is documented in *Science-Fiction Studies,* July 1977 ff.)

A Polish reviewer has remarked that Lem is not interested in literature *per se* at all; his main interest is in the structure of the world, not the structure of the literary work. Lem is more interested in intellectual problems than in their literary expression. He has no patience with the notion of fiction as entertainment or art for art's sake, and fiction without intellectual problems bores him. He is filled with curiosity about what is not yet known. For him, science fiction is a laboratory for trying out experiments in new ways of thinking; it should be a spearhead of cognition. It should attempt what hasn't been thought or done before.

These goals are of course impossible to achieve, let alone in a literature of mass entertainment. Lem sees science fiction as literature with great potential—a potential that naïve apologists often claim has already been achieved—and he is all the more disappointed that it falls so far short of his expectations. He complains that it is only a rehash of old myths and fairy tales, that it avoids all kinds of real problems, and that it resorts to narrative patterns of primitive adventure literature, which are wholly inadequate to express what is claimed for science fiction. For him, science fiction plays "empty games"—the tired old vaudeville of time travel, robots, supermen, mutants, extrasensory perception, and the rest.

From this disappointment comes the polemical acerbity of Lem's writings on science fiction. He believes in old-fashioned cultural and intellectual virtues, which he sees threatened by the onslaught of mass culture. Science fiction is a traitor to those values; even worse, it often claims to possess those values when it does not, quite unlike more modest forms of popular fiction. One of Lem's recurrent nightmares is the flood of information whose sheer volume makes it extremely difficult, if not impossible, to find the few good works in the mass of the bad. Such a leveling effect he also attributes to structuralism, one of his main targets.

It is Lem's concern for the real world and its cultural heritage that explains the sharpness of his tone, for he writes in a tradition where cultural values matter. Lem does not have to denigrate "Western science fiction" in order to ingratiate himself with the Polish authorities, as one of the sillier opinions once current among American science-fiction writers had it. Nor does it follow, from the fact that Lem's criticisms in *Fantastyka i futurologia* deal mostly with American and English science fiction, that he likes the Soviet variety any better. It is only that it would be more difficult for a citizen of Poland to write serious criticism about Soviet science fiction, not because it couldn't be published, but rather because objective discussion of the better works would be fraught with dangers for the writers discussed. Nevertheless, the present collection includes an analysis of one of the most brilliant examples of modern Soviet science fiction, *Roadside Picnic,* by the Strugatsky brothers, originally published as an afterword to a Polish edition in a series entitled "Lem Recommends." (The series included Philip K. Dick's *Ubik*— Lem's afterword is also printed here—and short story collections by M. R. James and by the Polish weird fiction writer Stefan Grabiński, as well as Ursula K. Le Guin's *Wizard of Earthsea.*)

All the essays in this volume have been published before, and the selection presented here was made entirely from existing translations. The magazines in which they first appeared range from science-fiction fanzines to *Science-Fiction Studies* and *The New Yorker.* Some were written originally in German; two came into English via Hungarian and German, respectively. Of some essays, such as the one on Borges, there is no Polish version; one, "Todorov's Fantastic Theory of Literature," was originally written in German, later rewritten in Polish by the author himself, and then translated from Polish to English. The autobiographical essay was commissioned by a Chicago publisher for a series on contemporary authors. It was written by Lem in German and published in my translation. A somewhat different version appeared in *The New Yorker,* as "Chance and Order."

Despite the mixed origins of the essays and the many hands involved in giving them their final shape, I think that this collection has a remarkable unity and can serve as a useful introduction to Lem's nonfiction and to his ideas on science fiction and fantasy. It should contribute to a better understanding of Lem's unique fiction, so much different in scope from other science fiction even when it uses the same forms.

Special thanks are due to all the people who first published these essays, especially to Bruce Gillespie, whose *Science Fiction Commentary* was in its day one of the liveliest and most open of the many fan magazines in the science-fiction field.

—*Franz Rottensteiner*

REFLECTIONS ON MY LIFE

As I write this autobiographical essay, I am aware of two opposed principles that guide my pen. One of those two extremes is chance; the other is the order that gives shape to life. Can all the factors that were responsible for my coming into the world and enabled me, although threatened by death many times, to survive unscathed in order finally to become a writer—moreover, one who ceaselessly strives to reconcile contradictory elements of realism and fantasy—be regarded only as the result of long chains of chance? Or was there some specific predetermination involved, not in the form of some supernatural *moira,* not quite crystallized into fate when I was in my cradle but in a budding form laid down in me—that is to say, in my genetic inheritance was there a kind of predestiny befitting an agnostic and empiricist?

That chance played a role in my life is undeniable. In the First World War, when the fortress of Przemyśl fell, in 1915, my father, Samuel Lem, a physician in the Austro-Hungarian Army, was taken prisoner by the Russians, and was able to return to Lemberg (now Lvov), his native city, only after nearly five years, in the wake of the chaos of the Russian Revolution. I know from the stories he told us that on at least one occasion he was to be shot by the Reds on the spot for being an officer (and therefore a class enemy). He owed his life to the fact that when he was being led to his execution in a small Ukrainian city he was noticed and recognized from the sidewalk by a Jewish barber from Lemberg who used to shave the military commander in that city and for this reason had free access to him. The barber interceded for my father (who was then not yet my father), and he was allowed to go free, and was able to return to Lemberg and to his fiancée. (This story, made more complex for aesthetic reasons, is to be found in one of the fictitious reviews—of "De Impossibilitate Vitae," by Cezar Kouska—in my book *A Perfect Vacuum.*) In this instance, chance was fate incarnate, for if the barber had happened to pass through that street a minute later my father would have been irrevocably doomed. I heard the tale from him when I was a little boy, at a time when I was unable to think in abstract terms (I may have been ten), and was thus unable to consider the respective merits of the categories of chance and fate.

My father went on to become a respected and rather

wealthy physician (a laryngologist) in Lvov. I was born there in 1921. In the rather poor country that Poland was before the Second World War, I lacked nothing. I had a French governess and no end of toys, and for me the world I grew into was something final and stable. But, if that was the case, why did I as a child delight in solitude, and make up the rather curious game that I have described in another book—the novel *The High Castle,* a book about my early childhood. My game was to transport myself into fictitious worlds, but I did not invent or imagine them in a direct way. Rather, I fabricated masses of important documents when I was in high school in Lvov: certificates; passports; diplomas that conferred upon me riches, high social standing, and secret power, or "full power of authority," without any limit whatsoever; and permits and coded proofs and cryptograms testifying to the highest rank—all in some other place, in a country not to be found on any map. Did I feel insecure in some way? Threatened? Did this game perhaps spring from some unconscious feeling of danger? I know nothing of any such cause.

I was a good student. Some years after the war, I learned from an older man who had held some position or other in the prewar Polish educational system that when the IQs of all high-school students were tested—it must have been around 1936 or 1937—mine was over 180, and I was said to have been, in the words of that man, the most intelligent child in southern Poland. (I myself suspected nothing of this sort at the time of the test, for the results were not made known to us.) But this high IQ certainly was of no help in surviving the occupation of the *General-gouvernement* (to which administrative unit Poland had been reduced by the Germans). During that period, I learned in a very personal, practical way that I was no "Aryan." I knew that my ancestors were Jews, but I knew nothing of the Mosaic faith and, regrettably, nothing at all of Jewish culture. So it was, strictly speaking, only the Nazi legislation that brought home to me the realization that I had Jewish blood in my veins. We succeeded in evading imprisonment in the ghetto, however. With false papers, my parents and I survived that ordeal.

But, to return to my childhood in prewar Poland, my first reading matter was of a rather curious nature. It was my father's anatomy books and medical texts, in which I browsed when I was still hardly able to read, and I understood them all the less since my father's professional books were in German or in French. Only the fiction in his library was in Polish. Pictures of skeletons, of neatly dissected human skulls, of human brains precisely sketched in many colors, of intestines in preserved condition and embellished with magic-sounding Latin names provided my earliest contacts with the world of books. Hunting through my father's library was, of course, strictly forbidden to me, and it attracted me precisely because it was forbidden and mysterious. I must not forget to mention the actual human bone that was kept behind the glass doors of my father's bookcase. It was a skull bone—*os temporale*—that had been removed during a trepanation; perhaps it was a relic from the time when my father was studying medicine. I held this bone, without any particular feelings, several times in my hands. (I had to steal my father's key to be able to do this.) I knew what it was, but I wasn't frightened by it. I only wondered about it in a certain way. Its surroundings—the rows of big tomes of medical textbooks—appeared quite natural to me, for a child, lacking any real yardstick, is unable to differentiate between the banal or commonplace and the unusual. That bone—or, rather, its fictional counterpart—is to be found in another novel of mine, *Memoirs Found in a Bathtub*. In this book, the bone became a whole skull, cleanly dissected from a corpse, that was kept by a doctor in a ward—one of the many stations in the hero's odyssey through a labyrinthine building. A complete skull like this was owned by my uncle, my mother's brother, who was also a physician. He was murdered two days after the Wehrmacht marched into Lvov. At that time, several non-Jewish Poles were also killed—mostly university professors—and Tadeusz Boy-Żeleński, one of the best-known Polish writers. They were taken from their apartments during the night and shot.

Now, then, what objective, extrinsic connection—i.e., not one imagined by me and consisting solely of associations—could there be between a little boy's fascination with the parts of a human skeleton and the era of the Holocaust? Was this apparently significant and fitting omen a matter of chains of chance, purely of coincidence? In my opinion, it was. I do not believe in manifest destiny or predetermination. In lieu of a preestablished harmony, I can well imagine (upon the basis of the experiences of my life) a preestablished disharmony, ending in chaos and madness. In any case, my childhood was certainly peaceful and Arcadian—especially when compared with what happened in the following years.

I grew into a bookworm, and read everything that fell

into my hands: the great national poems, novels, popular-science books. (I still remember that a book of the kind that my father gave me as a gift sold for seventy zlotys—the price was written inside—and that was a fortune in those days; for seventy zlotys you could buy a whole suit. My father spoiled me.) I also—I can still remember it—looked with keen interest at the male and female genitalia reproduced in my father's anatomy books. The female pubis struck me especially—as something spiderlike, not quite nauseating but certainly something that could hardly have a connection with erotic feelings. I believe that I was later, during my adolescence, sexually quite normal. But since my subsequent studies in medicine included gynecology, and since I was, for a month, an obstetrician in a hospital, I associate the pornography of today not with sexual longing and with copulative lust but with the anatomical pictures in the tomes of my father, and with my own gynecological examinations. The thought that a male may be highly excited by the mere sight of female genitalia strikes me as very peculiar. I happen to know perfectly well that this is a case of libido—of the instincts built into our senses and programmed by evolution—but the desire for sex without love strikes me as something comparable to an irresistible urge to eat salt and pepper by the spoonful because dishes without salt and pepper lack full flavor. I feel no repulsion but no attraction, either, as long as there is no specific erotic bond of the kind that is called "love."

As an eight-year-old boy, I fell in love with a girl. I never uttered as much as a word to the girl, but I observed her often in a public garden near our house. The girl had no inkling of my feelings, and most probably never even noticed me. It was a burning, long-lasting love affair dissected, as it were, from all actual circumstances—even from the sphere of any kind of wishful thinking. I was not interested in becoming her friend. My emotions were restricted to worshiping her from afar; aside from that, there was absolutely nothing. May the psychoanalysts make what they will of these feelings of a small boy. I do not comment further on them, because I am of the opinion that such an episode can be interpreted in any way one chooses.

At the beginning, I mentioned the opposites of chance and order, of coincidence and predestination. Only as I wrote the book *The High Castle* did the thought cross my mind that my fate—my profession as a writer—was already budding in me when I looked at skeletons, galaxies in astrophysical tomes, pictures of reconstructions of the monstrous extinct saurians of the Mesozoic, and many-colored human brains in anatomical handbooks. Perhaps these external circumstances—these impulses and sensuous impressions—helped to shape my sensibility. But that is only speculation.

I not only imagined fantastic kingdoms and domains but also made inventions and mentally created prehistoric animals unheard of in paleontology. For instance, I dreamed up an aircraft shaped like a giant concave mirror, with a boiler situated in the focus. The circumference of the mirror was studded with turbines and rotors to provide lift, as in a helicopter, and the energy for all that was to be derived from solar radiation. This unwieldy monstrosity was supposed to fly very high, far above the clouds, and, of course, only during daytime. And I invented what had already existed for a long time without my knowing it: the differential gear. I also drew many funny things in my thick copybooks, including a bicycle on which one rode moving up and down, as on a horse. Recently, I saw something like this imaginary bicycle somewhere—it may have been in the English periodical *New Scientist,* but I am not quite sure.

I think it is significant that I never bothered to show my designs to other people; indeed, I kept them all secret, both from my parents and from my fellow pupils, but I have no idea why I acted in this way. Perhaps it was because of a childish affection for the mysterious. The same was the case with my "passports"—certificates and permits that, for instance, allowed one to enter subterranean treasure troves. I suppose also that I was afraid to be laughed at, for, although I knew that these things were only a game, I played it with great seriousness. I divulged something of this childhood world in the book that I have already mentioned, *The High Castle,* but it contains only a small part of my memories. Why only a small part? I can answer such a question at least partly. First, in *The High Castle* I wanted to transport myself back into the child that I had been, and to comment on childhood as little as possible from the position of the adult. Second, during its gestation period the book generated a specific normative aesthetic similar to a self-organizing process, and there were certain memories that would appear as dissonances in this canon. It was not the case that I intended to hide certain things because of, say, a feeling of guilt or of shame but, rather, that there were memories that would not fit into the pattern that I presented as my childhood. I wanted—something impossible to attain—to extract the essence of my childhood, in its pure form, from my whole life: to peel away, as it were, the overlying strata of war, of mass murder and extermination, of the nights in the shelters during air raids, of an existence under a false identity, of hide-and-seek, of all the dangers, as if they had never existed. For, indeed, nothing of this had existed when I was a child, or even a sixteen-year-old high-school boy. I gave an indication of these exclusions in the novel itself. I do not remember exactly where, but I signaled that I had to or wanted to keep certain matters out by dropping a parenthetical remark that every human being is able to write several strikingly different autobiographies, according to the viewpoint chosen and the principle of selection.

The meaning of the categories of order and chance for human life was impressed upon me during the war years in a purely practical, instinctual manner; I resembled more a hunted animal than a thinking human being. I was able to learn from hard experience that the difference between life and death depended upon minuscule, seemingly unimportant things and the smallest of decisions: whether one chose this or that street for going to work; whether one visited a friend at one o'clock or twenty minutes later; whether one found a door open or closed. I cannot claim that in following my instinct for self-preservation I always employed a minimax strategy of extreme cautiousness. To the contrary, I exposed myself to danger several times—occasionally when I thought it necessary but in some cases through mere thoughtlessness, or even stupidity. So that today, when I think of such idiotically reckless patterns of behavior, I still feel wonder, mingled with bewilderment, about why I acted as I did. To steal ammunition from the so-called *Beutepark der Luftwaffe* (the depot where the German Air Force stored its loot) in Lvov and to turn it over to somebody totally unknown to me—somebody of whom I knew only that he was a member of the Resistance—I considered to be my duty. (I was in a position to do so since, as an employee of a German company, I had access to this depot.) But when I was instructed to transport something—a gun, in this case— from one place to another just before curfew, and was told, strictly, not to use the tram (I was supposed to walk), it happened that I nevertheless disobeyed the order and climbed onto the footboard of a tram, and that a "Black One"—a Ukrainian policeman who was a member of the auxiliary police of the German occupational forces— jumped onto the footboard behind me and put his arm around me to reach for the door handle. It could have meant an ill end for me if the policeman had felt the gun. My act was insubordination, thoughtlessness, and folly all in one, but I did it anyway. Was it a challenge to fate, or only foolhardiness? Up to this day, I am not sure. (I am better able to understand why I visited the ghetto several times—risky though this was—when it was open to visitors. I had friends there. As far as I know, all, or nearly all, of them were transported to the gas chambers of Belzec in the fall of 1942.)

At this point, the question arises whether what I have reported so far is relevant at all, in the sense of having any direct, causal relationship to my profession as a writer, or to the kind of writing I have done—excluding, of course, autobiographical works like *The High Castle*. I believe that such a causal relationship exists—that it isn't mere chance that I attribute in my work such a prominent role to chance as the shaper of human destiny. I have lived in radically different social systems. Not only have I experienced the huge differences in poor but independent, capitalist (if one must call it that) prewar Poland, the Pax Sovietica in the years 1939–41, the German occupation, the return of the Red Army, and the postwar years in a quite different Poland, but at the same time I have also come to understand the fragility that all systems have in common, and I have learned how human beings behave under extreme conditions—how their behavior when they are under enormous pressure is almost impossible to predict.

I remember well my feelings when I read *Mr. Sammler's Planet,* by Saul Bellow. Now, I thought that book very good—so good that I have read it several times. Indeed. But most of the things that Mr. Bellow attributed to his hero, Mr. Sammler, in recounting his experiences in a Poland occupied by the Germans, didn't sound quite right to me. The skilled novelist must have done careful research before starting on the novel, and he made only one small mistake—giving a Polish maid a name that isn't Polish. This error could have been corrected by a stroke of the pen. What didn't seem right was the "aura"—the indescribable "something" that can be expressed in language perhaps only if one has experienced in person the specific situation that is to be described. The problem in the novel is not the unlikeliness of specific events. The most unlikely and incredible things did happen then. It is, rather, the total impression that evokes in me the feeling that Bellow learned of such events from hearsay, and was in the situation of a researcher who receives the individual parts of a specimen packaged in separate crates and then tries to put them together. It is as if oxygen, nitrogen, and water vapor and the fragrance of flowers were to be mixed in such a way as to evoke and bring to life the specific mood of a certain part of a forest at a certain morning hour. I do not know whether something like this would be totally impossible, but it would surely be difficult as hell. There is something wrong in *Mr. Sammler's Planet;* some tiny inaccuracy got mixed into the compound. Those days have pulverized and exploded all narrative conventions that had previously been used in literature. The unfathomable futility of human life under the sway of mass murder cannot be conveyed by literary techniques in which individuals or small groups of persons form the core of the narrative. It is, perhaps, as if somebody tried by providing the most exact description of the molecules of which the body of Marilyn Monroe was composed to convey a full impression of her. That *would* be impossible. I do not know, of course, whether this sort of narrative inadequacy was the reason that I started writing science fiction, but I suppose—and this is a somewhat daring statement—that I began writing science fiction because it deals with human beings as a species (or, rather, with all possible species of intelligent beings, one of which happens to be the human species). At least, it *should* deal with the whole species, and not just with specific individuals, be they saints or monsters.

It is likely that, after my beginner's attempts—that is to say, after my first science-fiction novels—I revolted for the same reason of narrative limitations against the paradigms of the genre as they developed and became fossilized in the United States. As long as I didn't know current science fiction—and I didn't know it for a long time, because up to 1956 or 1957 it was almost impossible to get foreign books in Poland—I believed that it had to be a further development of the starting position taken by H. G. Wells in *The War of the Worlds*. It was he who climbed into a general's position, from which it was possible to survey the whole human species in an extreme situation. He anticipated a future filled with disasters, and I must admit that he was correct. During the war, when I read his novel several times, I was able to confirm his understanding of human psychology.

Today, I am of the opinion that my earliest science-fiction novels are devoid of any value (regardless of the fact that they had large editions everywhere and made me world-famous). I wrote these novels—for instance, *The Astronauts,* which was published in 1951, and was about an expedition to the planet Venus from a simplistically utopian Earth—for reasons that I can still understand today, although in their plots and in the kind of world they depicted they were contrary to all my experience of life at the time. In these books, the evil world of reality was supposed to have suffered a sea change into a good one. In the postwar years, there seemed to be only this choice —between hope and despair, between a historically untenable optimism and a well-justified skepticism that was easily apt to turn into nihilism. Of course, I wanted to embrace optimism and hope!

However, my very first novel was a realistic one, which

I wrote perhaps in order to rid myself of the weight of my war memories—to expel them like pus. But perhaps I wrote this book also in order not to forget; the one motive could well go together with the other. The novel is called *The Hospital of Transfiguration,* and it is about the fight of the staff of a hospital for the insane to save the inmates from being killed by the German occupiers. One German reviewer ventured the opinion that it was a kind of sequel to Thomas Mann's *The Magic Mountain.* What was in Mann only a portent—only the distant hint of a then nearly invisible lightning, since the horrors to come were still hidden behind the horizon of the times—proves to be in my novel the final circle of Hell, the logical outcome of the predicted "decline of the West" in the mass exterminations. The village, the hospital for the mentally ill, the professional staff: none of the places and characters ever existed; they are all my invention. But mentally ill persons —and many others—were indeed murdered by the thousands in occupied Poland. I wrote *The Hospital of Transfiguration* in 1948, my last year as a student. It could not appear until 1955, however, since it didn't conform to the then already reigning standards of Socialist Realism. In the meantime, I was, as I can say without exaggeration, very busy.

In 1946, we—my father, my mother, and I—moved from Lvov to Kraków, having lost all our possessions in the course of the war. My father, who was seventy-one years old, was forced, because of these reverses, to work in a hospital; there was no possibility that he could set up his own practice. We all lived in a single room in Kraków, and my father didn't have the means to buy his own equipment. Purely by chance, I learned how I could financially help our family: I wrote several long stories for a weekly dime-novel series that featured a complete story in each issue. Considered as thrillers, they weren't so bad. Aside from that, I wrote poems; they appeared in *Tygodnik Powszechny,* the Krakovian Catholic weekly. And two novellas—not science fiction proper but on the margin of fantasy—plus some odds and ends in various publications. But I did not take my writing very seriously.

In 1947, at the age of twenty-six, I became a junior research assistant for an organization called Konwersatorium Naukoznawcze (the Circle for the Science of Science), founded by Dr. Mieczysław Choynowski. To him I presented my most dearly held works: a theory of brain functions invented by me, and a philosophical treatise. He called both nonsense but took me under his tutelage. Thus, I was forced to read logic textbooks, scientific methodology, psychology, psychometrics (the theory of psychological testing), the history of natural science, and many other things. Although it was apparent that I couldn't read English, I had to do the best I could with English-language books. These books proved so interesting that I had to crack them, dictionary in hand, as Champollion cracked his hieroglyphs. Since I had learned French at home and Latin and German in school, and had picked up some Russian, I somehow managed to get along. But to this day I can understand only written English. I can neither speak the language nor understand it when it is spoken. For the monthly *Życie Nauki (The Life of Science),* I compiled surveys of scientific periodicals from the standpoint of the science of science. By doing so, I became involved in the wretched Lysenko affair, for in my survey I synopsized the controversy between him and the Soviet geneticists in what an official report from the ministry in charge of Polish universities called "a tendentious manner." I held Lysenko's doctrine of the inheritance of acquired characteristics to be ridiculous, and I was proved right after several years, but my taking this position had rather painful consequences for our monthly. Something similar happened a little later, when I perceived in Norbert Wiener's and Claude E. Shannon's cybernetics a new era not just for technological progress but also for the whole of civilization. At that time, cybernetics was considered in our country to be a fallacious pseudoscience.

In those years, I was particularly well informed about the latest developments in the various sciences, for the Krakovian circle functioned as a kind of clearinghouse for scientific literature from the United States (and, to some extent, from Canada) coming in to all the Polish universities. From the book parcels received I could borrow all the works that stirred my interest, including Wiener's *The Human Use of Human Beings.* At night, I read everything voraciously, so that I could pass on the books as soon as possible to the people who were supposed to get them. On the basis of this reading, I wrote those of my novels that I can still acknowledge without shame—*Eden* (1959), *Solaris* (1961), *The Invincible* (1964), etc. They incorporate cognitive problems in fictions that do not oversimplify the world, as did my earliest, naïve science-fiction novels.

My father died in 1954, and toward the end of the fifties I was able to acquire for us—myself and my wife—a small house on the southern outskirts of Kraków, which we still have. (Close to this house, a larger house, in a larger garden, is in the process of being built for us as I write these words.) In the late sixties, I first made contact with my future literary agent and kindred spirit, Franz Rotten-steiner, from Vienna. Both of us were then writing many critical, often polemical essays for Anglo-American science-fiction "fanzines" (i.e., the amateur magazines published by the aficionados of science fiction), mostly for Bruce Gillespie's Australian *Science Fiction Commentary*; that resulted in a certain popularity for both of us, even if it was of a negative sort, in the science-fiction ghetto. Today, I am of the opinion that we wasted our efforts. In the beginning, it was totally incomprehensible to me why so many authors were erecting, *viribus unitis,* a common prison for science fiction. I believed that, according to the law of large numbers alone, there had to be among so many a considerable group at the top, as far as both writing abilities and scientific qualifications were concerned. (For me, the scientific ignorance of most American science-fiction writers was as inexplicable as the abominable literary quality of their output.) I was in error, but it took me a very long time to recognize it.

As a reader of science fiction, I expected something like what is called, in the evolutionary processes of nature, "speciation"—a new animal species generating a diverging, fanlike radiation of other new species. In my ignorance, I thought that the time of Verne, Wells, and Stapledon was the beginning, but not the beginning of the decline, of the sovereign individuality of the author. Each of these men created something not only radically new for their time but also quite different from what the others created. They all had enormous room for maneuvering in the field of speculation, because the field had only recently been opened up and was still empty of both writers and books. Each of them entered the no man's land from a different direction and made some particular province of this *terra incognita* his own. Their successors, on the other hand, had to compromise more and more with the crowd. They were forced to become like ants in an ant hill, or industrious bees, each of which is indeed building a different cell in the honeycomb but whose cells are all similar. Such is the law of mass production. Thus, the distance between individual works of science fiction has not grown greater, as I erroneously expected, but has shrunk. The very thought that a Wells or a Stapledon could have written, alternately, visionary fantasies and typical mysteries strikes me as absurd. For the next generation of writers, however, this was something quite normal. Wells and Stapledon are comparable to the people who invented chess and draughts. They discovered new rules for games, and their successors have applied these rules with only smaller or larger variations. The sources of innovation have gradually become depleted; the thematic clusters have become fossilized. Hybrids have arisen (science fantasy), and the patterns and schemata of the literary form have been applied in a mechanical and ready-made way.

To create something radically new, it was necessary to advance into another field of possibilities. I believe that in the first period of my career I wrote purely secondary things. In the second period *(Solaris, The Invincible)*, I reached the borders of a field that was already nearly completely mapped. In the third period—when I wrote, for example, reviews of nonexistent books and forewords to works that, as I put it, ironically, in an interview, would be published "sometime in the future but that do not exist yet"—I left the fields already exploited and broke new ground. This idea is best explained by a specific example. A few years ago, I wrote a small book entitled *Provocation.* It is a review of a fictitious two-volume tome ascribed to a nonexistent German historian and anthropologist, whom I call Aspernicus. The first volume is titled *Die Endlösung als Erlösung* (The Final Solution Considered as Redemption), the second *Fremdkörper Tod* (Foreign Body Death). The whole thing is a unique historico-philosophical hypothesis about the as yet unrecognized roots of the Holocaust, and the role that death, especially mass death, has played in the cultures of all times up to the present day. The literary quality of my fictitious criticism (which is rather long, or it wouldn't have filled even a small book) is beside the point here. What counts is the fact that there were professional historians who took my fancy for the review of a real book, as is attested to by attempts on the part of some of them to get hold of the book. To my mind, *Provocation,* too, is a kind of science fiction; I am trying not to limit the meaning of the name of this category of writing but, rather, to expand it.

Nothing I've ever written was planned in an abstract form right from the start, to be embodied later in literary form. Nor can I claim that it was my intention to find other fields for development—that I set out with the intention of seeking them out for my imagination. But I can say something about the conception of an idea, the gravid state, the pains of giving birth, though I do not know the genetic make-up of the embryo or know how it is transformed into a phenotype—the finished work. Here, in the realm of the "embryogenesis" of my writing, considerable differences have developed in the course of some thirty-six years.

My earliest novels (which I acknowledge as my own only with some discomfort) I planned and constructed according to a complete design. I wrote the novels in the *Solaris* group in a similar manner, which I myself cannot explain. The terminology of birth that I have used above may sound inappropriate, but it is somewhat apt. I am still able to point to passages in *Solaris* and *Return from the Stars* where I found myself, during the writing process, in the position of a reader. When Kelvin, the narrator of *Solaris,* arrives at the station hovering over the planet Solaris and finds it empty of human beings, and when he starts his search for the crew, and encounters the scientist Snow, who goes into a state of panic when he sees Kelvin, I had no idea why nobody had expected his arrival or why Snow behaved in this peculiar manner; indeed, I had no idea at all that some "living ocean" would cover the whole planet. All this was divulged to me in the same manner that it becomes clear to the reader in the course of reading the book—with the sole difference that it was I who created the novel. And in *Return from the Stars* I faced a wall when the returning astronaut frightens one of the first women he meets, and then the word "betrization" is used: that's the treatment that human beings have undergone in the future world to rid them of their aggressive impulses. I didn't know at first exactly what the word should mean, but I knew that there must be some unbridgeable difference between the civilization that the man left when he flew to the stars and the one that he found upon his return. The metaphor that takes its terms from the lexicon of embryology is thus not nonsense, for a woman who is with child knows that she carries an embryo, but she has no idea how the embryo is transformed from an ovum into a child. Considering myself to be a rationalist, I dislike such confessions, and I should prefer to be able to say that I knew everything I was doing—or, at least, a good deal of it—beforehand, and that I planned and designed it, but *amicus Plato, sed magis amica veritas.*

Nevertheless, something can be said about my creative method. First, there is no positive correlation between the spontaneity of my writing and the quality of the resulting work. I gave birth to *Solaris* and *Return from the Stars* in a similar manner, but I think that *Solaris* is a good book and *Return from the Stars* a poor one, because in the latter the underlying problems of social evil and its elimination are treated in a manner that is too primitive, too unlikely, and perhaps even false. (Even if the evil done to others with full intent could be suppressed pharmacologically—the book's main premise—no chemical or other influence upon the brain could cause the unintended evil effects of all social dependencies, conflicts, and contradictions to disappear from the world, in the same manner that an insecticide can eliminate vermin.) Second, creative spontaneity is not a guarantee that there will be sure development of a whole narrative—i.e., a plot that can be finished without applying force. I have had to put more stories aside unfinished or drop them into the wastebasket than

I have been able to submit to publishers. Third, this process of writing, which is characterized by the signs of a creation by trial and error, has always been arrested by blocks and blind alleys that forced me to retreat; sometimes there has even been a "burning out" of the raw materials—the manifold resources necessary for further growth—stored somewhere in my skull. I was not able to finish *Solaris* for a full year, and could do it then only because I learned suddenly—from myself—how the last chapter had to be. (And then I could only wonder why I hadn't recognized it from the beginning.) And, fourth, even what I wrote spontaneously never received its final shape in the first thrust of work. I have never written a larger work (it is different with short stories) in a "linear" way right to the end in one sweep; rather, in the pauses between writing sessions—it is for purely physiological reasons impossible to sit at the typewriter all the time—I had new ideas that enriched what was already finished or was to be written soon; changed it; and complicated it with some new turn or complexity of plot.

Practical experience—the result of wrestling with my writing over the years—has taught me never to force what I am working on if it has not ripened at least partly but, rather, to let it rest for some time (which may amount to periods of months, or even years) and let the thing mill around in my head. (A gravid woman knows that an early birth bodes nothing good.) This situation has put me on the horns of a dilemma, however, for, like nearly all writers, I often try to invent excuses for not writing. As is well known, laziness is one of the main barriers hindering everyone in his work. If I waited until I carried something in its definite form around in my head, I would never create anything.

My method of creating (which I should like to call, rather, my behavior as a writer) has changed during the years, if only very slowly. I have learned to avoid the pure spontaneity of beginnings which motivated me to write something even when I had not the slightest idea what would come of the thing—its plot, its problems, its characters, etc.—because the instances in which I was unable to finish what I had begun were on the increase. Perhaps the imaginative space that was given me became gradually emptied, like a territory rich in oil, from which the black gold at first fountains in the air everywhere in geysers, no matter where one begins to drill; after some time, one has to use ever more complicated tricks and apply pressure to drive the remaining reserves up to the surface. The center of gravity of my work, then, gradually shifted in the direction of the gaining of a basic idea, a conception, an imaginative notion. I ceased to sit down at my typewriter whenever I had a quite small but ready beginning; instead, I started to produce an increasing number of notes, fictitious encyclopedias, and small additional ideas, and this has finally led to the things I am doing now. I try to get to know the "world" to be created by me by writing the literature specific to it, but not whole shelves of reference works of the sociology and the cosmology of some thirtieth century, not the fictitious minutes of scientific expeditions or other types of literature that express a *Zeitgeist,* the spirit of a time and a world, alien to us. After all, this would be an endeavor impossible to accomplish during the short life span of a human being. Nor do I now do what began in the first place rather as a joke—write criticism in the form of the reviews of nonexistent books or forewords to them *(A Perfect Vacuum, Imaginary Magnitude).* I do not publish these things any longer but use them to create my own knowledge of another world, a knowledge entirely subservient to my literary program—in other words, to sketch a rough outline that will be filled in later. I surround myself, so to speak, with the literature of a future, another world, a civilization with a library that is its product, its picture, its mirror image. I write only brief synopses or, again, critical reviews of sociological treatises, scientific papers, and technical reference works, and I describe technologies that have taken the place of literature after its final death, just as television has made obsolete the *cinématographe* of Lumière, and three-dimensional television will make obsolete the TV sets of today. There are also historicophilosophical papers, "encyclopedias of alien civilizations" and their military strategies—all of them, of course, in a kind of shorthand, or I would need the longevity of a Methuselah to create them. It may well be that I will publish something out of this "library for a given purpose" independently of the work for which it served as a frame and a source of information.[1]

---

1. Michel Butor once expressed the opinion that a team of science-fiction writers should cooperate in the construction of a fictitious world, because such an undertaking is beyond the powers of any single individual. (This was supposed to explain the poor quality, the one-dimensionality of the existing science fiction.) I did not take those words of Butor's seriously when I read them. And yet I have, although many years later and by myself, tried to realize the basic essence of this idea as described above. And in Borges, too—in his "Tlön, Uqbar, Orbis Tertius"—you can read of a secret society that creates a fictitious world in all its particulars, with the intent of turning our world into the imagined one.

And where do I get all these facts, which I adorn with such enchanting titles as "The Trend of Dehumanization in Weapon Systems of the 21st Century" or "Comparative Culturology of Humanoid Civilizations"? In a certain sense, from my head; in another, not. I have invented several picturesque similes to illustrate for myself and others what my working method is like:

(1) A cow produces milk—that is certain—and the milk doesn't come from nothing. Just as a cow must eat grass in order to be able to produce milk, I have to read large amounts of genuine scientific literature of all kinds—i.e., literature not invented by me—and the final product, my writing, is as unlike the intellectual food as milk is unlike grass.

(2) Just as the ape in Wolfgang Köhler's psychological experiments wasn't able to reach a banana hanging very high, and made a scaffold from junk—boxes lying around, etc.—in order to be able to climb up to the banana, I have to build up in subsequent moves and attempts an informational "scaffold" that I must climb up to reach my goal.

(3) The last simile is somewhat drastic and may appear to be very primitive, but it nevertheless contains some grain of the truth. A water closet has a reservoir that must be filled, and when the lever or button is pushed all the water flushes down in one stream. Thereafter, the reservoir is empty for a time, and until it has been filled again no impatient pushing of the button or the lever will cause the small Niagara to flush forth again. As far as my work is concerned, this image is appropriate, in that if I did not keep enriching my fictitious library there would come a state of depletion, and after that I would not be able to get anything more out of my mind—my information store-

house. I wrote *A Perfect Vacuum*—it contains fifteen fictitious book reviews—nearly without a pause, and after that my reservoir was empty. Indeed, the comparison can be dragged in a little further. Just as, if you push the button of a toilet too soon, there will flush down only inadequate Niagaras, I can squeeze a little more from my head after the writing of a book like *A Perfect Vacuum*. But I will not be satisfied with the stuff gained this way, and I cast these remnants aside.

My working methods are additionally complicated and enriched by my having from time to time written quasi-scientific works that were not intended as scaffoldlike supports for fiction but meant seriously as independent books on the theory of literature (but they are along empirical lines that are alien to specialists in the humanities). And I have produced *Science Fiction and Futurology* (1970), which is an acerbic criticism and theory of science fiction; and skeptical futurology, like *Summa Technologiae* (1964), which doesn't amass many speculations about the wonderful or terrible things of the near future but, rather, attempts to pursue a few radical ideas to their utmost limits; and the *Dialogues* (1957), about the horizons and chances of cybernetics implicit in the system; and essays on various topics, such as *Biology and Values* (1968) and *Applied Cybernetics: An Example from the Field of Sociology*[2] (1971)—a discussion of the pathology of socialism.

---

2. I shall add the autobiographical element in my discursive writings to this enumeration. In brief, I am a disenchanted reformer of the world. My first novels concerned naïve utopias, because in them I was expressing a desire for a world as peaceful as that described in them, and they are bad, in the sense in which a vain and erroneous expecta-

Later, it turned out that several of the ideas that occurred to me during the writing of these works and that I used as hypotheses and examples—i.e., much of what I encountered on my chosen intellectual way during the process of writing—could also be put to good use in fiction. At first, this happened in a totally unconscious manner. I noticed it only when it was pointed out to me; that is, my critics discovered the similarities and were of the opinion that I oscillated with full consciousness between serious discussion and fantastic literature, when I myself was not aware of such a seesawing. Once my attention was drawn to this phenomenon, I sometimes browsed in my own books with an eye toward this possibility of exploitation or cross-fertilization.

In looking back, I see clearly that in my middle period as a writer I wrote fiction without any regard for the existence of some continuity between the imagined worlds and our world. In the worlds of *Solaris, Eden,* and *Return from the Stars,* there are no immediately obvious transitional stages that could connect these states of civilization with the obnoxious state of things on earth today. My later

---

tion is stupid. My monograph on science fiction and futurology is an expression of my disappointment with a fiction and a nonfiction that pretend to be scientific, when neither of them turns the attention of the reader in the direction in which the world is in fact moving. My *Philosophy of Chance* is a failed attempt to arrive at a theory of the literary work based on empiricism; it is successful inasmuch as I taught myself with the help of this book what factors cause the rise and the decline in the fortunes of literary works. My *Summa Technologiae,* on the other hand, is proof of the fact that I am not yet a despairing reformer of the world. For I do not believe that mankind is for all times a hopeless and incurable case.

work, on the other hand, shows marked signs of a turning toward our world; that is, my later fictions are attempts to establish such connections. I sometimes call this my inclination toward realism in science fiction. Most likely, such attempts, which to some extent have the unmistakable character of a retreat (as a renunciation of both utopia and dystopia, extremes that are either repugnant to me or leave me cold, just as is the case for a physician when he faces someone incurably ill), spring from the awareness that I must soon die, and from the resulting desire to satisfy, at least with hypotheses, my insatiable inquisitiveness about the far future of mankind and the cosmos. But that is only a guess; I wouldn't be able to prove it.

In response to a request to write his autobiography, Einstein emphasized not the historical circumstances of his life but, rather, his most beloved offspring—his theories—because they were the children of his mind. I am no Einstein, but in this respect I nevertheless resemble him, for I am of the opinion that the most important parts of my biography are my intellectual struggles. The rest, not mentioned so far, is of a purely anecdotal character.

In 1953, I married a young student of medicine. We have a son of fifteen, who likes my novels well enough but modern music—pop, rock and roll, the Beatles—his motorcycle, and the engines of automobiles perhaps even more. For many years now, I have not owned my books and my work; rather, I have become owned by them. I usually get up a short time before five in the morning and sit down to write: I am writing these words at six o'clock. I am unable now to work more than five or six hours a day without a pause. When I was younger, I could write as long as my stamina held out; the power of my intellect gave way only after my physical prowess had been exhausted. I write increasingly slowly—my self-criticism, the demands I put upon myself, have continued to grow —but I am still rather prolific. (I know this from the speed with which I have to throw away used-up typewriter ribbons.) Less and less of what comes into my mind I consider to be good enough to test as suitable subject matter by my method of trial and error. I still know as little about how and where my ideas are born as most writers do. I am also not of the opinion that I am one of the best exegetes of my own books—i.e., of the problems characteristic of them. I have written many books of which I haven't said a word here, among them *The Cyberiad*, the *Fables for Robots* (in *Mortal Engines*), and *The Star Diaries*, which on the generic map of literature are to be found in the provinces of the humorous—of satire, irony, and wit— with a touch of Swift and of dry, mischievous Voltairean misanthropy. As is well known, the great humorists were people who had been driven to despair and anger by the conduct of mankind. In this respect, I am one of those people.

I am probably both dissatisfied with everything that I have written and proud of it: I must be touched by arrogance, but I do not feel anything of it. I can notice it only in my behavior—in the way that I used to destroy all my manuscripts, in spite of many attempts and requests to get me to deposit these voluminous papers in a university or some other repository to preserve them for posterity. I have made up a striking explanation for this behavior. The pyramids were one of the wonders of the world only while there was no explanation of how they were erected. Very long, inclined planes, on which bands of workers hauled up the stone blocks, possibly on wooden cylinders, were leveled once the work was finished, and thus today the pyramids rise up in a lonely way among the shallow sand dunes of the desert. I try to level my inclined plane, my scaffolds and other means of construction, and to let stand only that of which I need not be ashamed.

I am not sure whether what I have confessed here is the pure truth, but I have tried to adhere to truth as well as I could.

*Translated from the German
by Franz Rottensteiner*

ON THE STRUCTURAL ANALYSIS OF
SCIENCE FICTION

In the early stages of literary development the different branches of literature, the genological types, are distinguished clearly and unmistakably. Only in the more advanced stages do we find hybridization. But since some cross-breedings are always forbidden, there exists a main law of literature that could be called "incest prohibition"; that is, the taboo of genological incest.

A literary work considered as a game has to be played out to the finish under the same rules with which it was begun. A game can be empty or meaningful. An empty game has only inner semantics, for it derives entirely from the relationships that obtain between the objects with which it is played. On a chessboard, for example, the king has its specific meanings within the rules of the play, but it has no reference outside the rules (i.e., it is nothing at all in relation to the world outside the confines of the chessboard). Literary games can never have so great a degree of semantic vacuum, for they are played with "natural language," which always has meanings oriented toward the world of real objects. Only with a language especially constructed to have no outward semantics, such as mathematics, is it possible to play empty games.

In any literary game there are rules of two kinds: those that realize outer semantic functions as the game unfolds and those that make the unfolding possible. "Fantastic" rules of the second kind—those that make the unfolding possible—are not necessarily felt as such even when they imply events that could not possibly occur in the real world. For example, the thoughts of a dying man are often detailed in quite realistic fiction even though it is impossible, therefore fantastic, to read the thoughts of a dying man out of his head and reproduce them in language. In such cases we simply have a convention, a tacit agreement between writer and reader—in a word, the specific rule of literary games that allows the use of nonrealistic means (e.g., thought reading) for the presentation of realistic happenings.

Literary games are complicated by the fact that the rules that realize outer semantic functions can be oriented in several directions. The main types of literary creation imply different ontologies. But you would be quite mistaken if you believed, for example, that the classical fairy tale has only its autonomous inner meanings and no relationship with the real world. If the real world did not exist, fairy tales would have no meaning. The events that occur in a myth or fairy tale are always semantically connected with what fate has decreed for the inhabitants of the depicted world, which means that the world of a myth or fairy tale is ontologically either *inimical* or *friendly* toward its inhabitants, never *neutral;* it is thus ontologically different from the real world, which may be here defined as consisting of a variety of objects and processes that lack intention, that have no meaning, no message, that wish us neither well nor ill, that are just there. The worlds of myth or fairy tale have been built either as traps or as happiness-giving universes. If a world without intention did not exist —that is, if the real world did not exist—it would be impossible for us to perceive the *differentia specifica,* the uniqueness, of the myth and fairy-tale worlds.

Literary works can have several semantic relationships at the same time. For fairy tales the inner meaning is derived from the contrast with the ontological properties of the real world, but for anti–fairy tales, such as those by Mark Twain in which the worst children live happily and only the good and well bred end fatally, the meaning is arrived at by turning the paradigm of the classical fairy tale upside down. In other words, the first referent of a semantic relationship need not be the real world but may instead be the typology of a well-known class of literary games. The rules of the basic game can be inverted, as they are in Mark Twain, and thus is created a new generation, a new set of rules—and a new kind of literary work.

In the twentieth century the evolution of mainstream literary rules has both allowed the author new liberties and simultaneously subjected him to new restrictions. This evolution is antinomical, as it were. In earlier times the author was permitted to claim all the attributes of God: nothing that concerned his hero could be hidden from him. But such rules had already lost their validity with Dostoevsky, and God-like omniscience with respect to the world he has created is now forbidden the author. The new restrictions are realistic in that as human beings we act only on the basis of *incomplete* information. The author is now one of us; he is not allowed to play God. At the same time, he is allowed to create inner worlds that need not necessarily be similar to the real world, but can instead show different kinds of deviation from it.

These new deviations are very important to the contemporary author. The worlds of myth and fairy tale also deviate from the real world, but individual authors do not invent the ways in which they do so: in writing a fairy tale you must accept certain axioms you haven't invented, or you won't write a fairy tale. In mainstream literature, however, you are now allowed to attribute pseudo-ontological qualities of your personal, private invention to the world you describe. Since all deviations of the described world from the real world necessarily have a *meaning,* the sum of all such deviations is (or should be) a coherent strategy or semantic *intention.*

Therefore we have two kinds of literary fantasy: "final" fantasy, as in fairy tales and science fiction, and "passing" fantasy, as in Kafka. In a science-fiction story, the presence of intelligent dinosaurs does not usually signal the presence of hidden meaning. The dinosaurs are, instead, meant to be admired as we would admire a giraffe in a zoological garden; that is, they are intended not as parts of an expressive semantic system, but only as parts of the empirical world. In *The Metamorphosis,* on the other hand, it is not intended that we should accept the transformation of human being into bug simply as a fantastic marvel, but, rather, that we should pass on to the recognition that Kafka has with objects and their deformations depicted a sociopsychological situation. Only the outer shell of this world is formed by the strange phenomena; the inner core has a solid nonfantastic meaning. Thus a story can depict the world as it is, or interpret the world (attribute values to it, judge it, call it names, laugh at it, etc.), or, in most cases, do both things at the same time.

If the depicted world is oriented positively toward man, it is the world of the classical fairy tale, in which physics is controlled by morality, for in a fairy tale there can be no physical accidents that result in anyone's death, no irreparable damage to the positive hero. If it is oriented negatively, it is the world of myth ("Do what you will, you'll still become guilty of killing your father and committing incest"). If it is neutral, it is the real world—the world that realism describes in its contemporary shape and that science fiction tries to describe at other points on the space-time continuum.

It is the premise of science fiction that anything shown shall in principle be interpretable empirically and rationally. In science fiction there can be no inexplicable marvels, no transcendences, no devils or demons—and the pattern of occurrences must be verisimilar.

And now we come near the rub, for what is meant by a verisimilar pattern of occurrences? Science-fiction authors try to blackmail us by calling upon the omnipotence of science and the infinity of the cosmos as a continuum. "Anything can happen" and therefore "anything that happens to occur to us" can be presented in science fiction.

But it is not true, even in a purely mathematical sense, that anything can happen, because there are infinities of quite different powers. Let us leave mathematics alone, however. Science fiction can be either "real science fiction" or "pseudo–science fiction." When it produces fantasy, like Kafka's, it is only pseudo–science fiction, because it then concentrates on the content to be signaled. What meaningful and total relationships obtain between the telegram "Mother died, funeral Monday" and the structure and function of the telegraphic apparatus? None. The apparatus merely enables us to transmit the message, which is also the case with semantically dense objects of a fantastic nature, such as the metamorphosis of man into bug, that nevertheless transmit a realistic communication.

If we were to change railway signals so that they ordered the stopping of trains in moments of danger not by blinking red lights but by pointing with stuffed dragons, we would be using fantastic objects as signals, but those objects would still have a real, nonfantastic function. The fact that there are no dragons has no relationship to the real purpose or method of signaling.

As in life we can solve real problems with the help of images of nonexistent beings, so in literature can we signal the existence of real problems with the help of prima facie impossible occurrences or objects. Even when the happenings it describes are totally impossible, a science-fiction work may still point out meaningful, indeed rational, problems. For example, the social, psychological, political, and economic problems of space travel may be depicted quite realistically in science fiction even though the *technological* parameters of the spaceships described are quite fantastic in the sense that it will for all eternity be impossible to build a spaceship with such parameters.

But what if everything in a science-fiction work is fantastic? What if not only the objects but also the problems have no chance of ever being realized, as when impossible time-travel machines are used to point out impossible time-travel paradoxes? In such cases science fiction is playing an empty game.

Since empty games have no hidden meaning, since they represent nothing and predict nothing, they have no relationship at all to the real world and can therefore please us only as logical puzzles, as paradoxes, as intellectual acrobatics. Their value is autonomous, because they lack all semantic reference; therefore they are worthwhile or worthless only as games. But how do we evaluate empty games? Simply by their formal qualities. They must contain a multitude of rules; they must be elegant, strict, witty, precise, and original. They must therefore show at least a minimum of complexity and an inner coherence; that is, it must be forbidden during the play to make any change in the rules that would make the play easier.

Nevertheless, ninety to ninety-eight percent of the empty games in science fiction are very primitive, very naïve one-parameter processes. They are almost always based on only one or two rules, and in most cases it is the rule of inversion that becomes their method of creation.

To write such a story you invert the members of a pair of linked concepts. For example, we think the human body quite beautiful, but in the eyes of an extraterrestrial we are all monsters: in Sheckley's "All the Things You Are" the odor of human beings is poisonous for extraterrestrials, and when they touch the skin of humans they get blisters, etc. What appears normal to us is abnormal to others— about half of Sheckley's stories are built on this principle. The simplest kind of inversion is a chance mistake. Such mistakes are great favorites in science fiction: something that doesn't belong in our time arrives here accidentally (a wrong time-mailing), etc.

Inversions are interesting only when the change is in a basic property of the world. Time-travel stories originated in that way: time, which is irreversible, acquired a reversible character. On the other hand, any inversion of a local kind is primitive (on earth humans are the highest biological species, on another planet humans are the cattle of intelligent dinosaurs; we consist of albumen, the aliens of silicon; etc.). Only a nonlocal inversion can have interesting consequences; we use language as an instrument of communication; any instrument can in principle be used for the good or bad of its inventor. Therefore the idea that language can be used as an instrument of enslavement, as in Delany's *Babel-17,* is interesting as an extension of the hypothesis that world view and conceptual apparatus are interdependent; i.e., because of the ontological character of the inversion.

The pregnancy of a Virgo Immaculata; the running of 100 meters in 0.1 seconds; the equation $2 \times 2 = 7$; the pan-psychism of all cosmic phenomena postulated by

Stapledon: these are four kinds of fantastic condition.

1. It is in principle possible, even empirically possible, to start embryogenesis in a virgin's egg; although empirically improbable today, this condition may acquire an empirical character in the future.

2. It will always be impossible for a man to run 100 meters in 0.1 seconds. For such a feat a man's body would have to be so totally reconstructed that he would no longer be a man of flesh and blood. Therefore a story based on the premise that a human being as a human being could run so fast would be a work of fantasy, not science fiction.

3. The product of 2 × 2 can never become 7. To generalize, it is impossible to realize any kind of logical impossibility. For example, it is logically impossible to give a logical proof for the existence or nonexistence of a god. It follows that any imaginative literature based on such a postulate is fantasy, not science fiction.

4. The pan-psychism of Stapledon is an ontological hypothesis. It can never be proved in the scientific sense: any transcendence that can be proved experimentally ceases to be a transcendence, for transcendence is by definition empirically unprovable. God reduced to empiricism is no longer God; the frontier between faith and knowledge can therefore never be annulled.

But when any of these conditions, or any condition of the same order, is described not to postulate its real existence, but only to interpret some content of a semantic character by means of such a condition used as a signal-object, then all such classificatory arguments lose their power.

What, therefore, is basically wrong in science fiction is

the abolition of differences that have a categorical character; the passing off of myths and fairy tales for quasi-scientific hypotheses or their consequences, and of the wishful dream or horror story as prediction; the postulation of the incommensurable as commensurable; the depiction of the accomplishment of possible tasks with means that have no empirical character; the pretense that insoluble problems (such as those of a logical *typus*) are soluble.

But why should we deem such procedures wrong when once upon a time myths, fairy tales, sagas, fables were highly valued as keys to all cosmic locks? It is the spirit of the times. When there is no cure for cancer, magic has the same value as chemistry: the two are wholly equal in that both are wholly worthless. But if there arises a realistic expectation of achieving a victory over cancer, at that moment the equality will dissolve, and the possible and workable will be separated from the impossible and unworkable. It is only when the existence of a rational science permits us to rule the phenomena in question that we can differentiate between wishful thinking and reality. When there is no source for such knowledge, all hypotheses, myths, and dreams are equal; but when such knowledge begins to accumulate, it is not interchangeable with anything else, because it involves not just isolated phenomena but the whole structure of reality. When you can only dream of space travel, it makes no difference what you use as technique: sailing ships, balloons, flying carpets or flying saucers. But when space travel becomes fact, you can no longer choose what pleases you rather than real methods.

The emergence of such necessities and restrictions often

goes unnoticed in science fiction. If scientific facts are not simplified to the point where they lose all validity, they are put into worlds categorically, ontologically different from the real world. Since science fiction portrays the future or the extraterrestrial, the worlds of science fiction necessarily deviate from the real world, and the ways in which they deviate are the core and meaning of the science-fiction creation. But what we usually find is not what may happen tomorrow but the forever impossible, not the real but the fairy-tale-like. The difference between the real world and the fantastic world arises stochastically, gradually, step by step. It is the same kind of process as that which turns a head full of hair into a bald head: if you lose a hundred, even a thousand hairs, you will not be bald; but when does balding begin—with the loss of 10,000 hairs or 10,950?

Since there are no humans that typify the total ideal average, the paradox of the balding head exists also in realistic fiction, but there at least we have a guide, an apparatus in our head that enables us to separate the likely from the unlikely. We lose this guide when reading portrayals of the future or of galactic empires. Science fiction profits from this paralysis of the reader's critical apparatus, because when it simplifies physical, psychological, social, economic, or anthropological occurrences, the falsifications thus produced are not immediately and unmistakably recognized as such. During the reading one feels instead a general disturbance; one is dissatisfied; but because one doesn't know how it should have been done, one is often unable to formulate a clear and pointed criticism.

If science fiction is something more than just fairy-tale fiction, it has the right to neglect the fairy-tale world and its rules. It is also not realism, and therefore has the right

to neglect the methods of realistic description. Its genological indefiniteness facilitates its existence, since it is sup-
posedly not subject to the whole range of the criteria by
which literary works are normally judged. It is not allegorical; but then it says that allegory is not its task: science
fiction and Kafka are two quite different fields of creation.
It is not realistic; but then, it is not a part of realistic
literature. The future? How often have science-fiction authors disclaimed any intention of making predictions! Finally, it is called the Myth of the Twenty-first Century.
But the ontological character of myth is antiempirical,
and though a technological civilization may have its
myths, it cannot itself embody a myth. For myth is an
interpretation, a *comparatio,* an explication, and first you
must have the object that is to be explicated. Science
fiction lives in but strives to emerge from this antinomical
state of being.

A quite general symptom of the sickness in science
fiction can be found by comparing the spirit in ordinary
literary circles to that in science-fiction circles. In the
literature of the contemporary scene there is today uncertainty, distrust of all traditional narrative techniques, dis-
satisfaction with newly created work, general unrest that
finds expression in ever new attempts and experiments; in
science fiction, on the other hand, there is general satisfaction, contentedness, pride; and the results of such comparisons must give us some food for thought.

I believe that the existence and continuation of the great
and radical changes effected in all fields of life by technological progress will lead science fiction into a crisis, which
is perhaps already beginning. It becomes more and more
apparent that the narrative structures of science fiction

deviate more and more from all real processes, having been used again and again since they were first introduced and having thus become frozen, fossilized paradigms. Science fiction involves the art of putting hypothetical premises into the very complicated stream of sociopsychological occurrences. Although this art once had its master in H. G. Wells, it has been forgotten and is now lost. But it can be learned again.

The quarrel between the orthodox and heterodox parts of the science-fiction fraternity is regrettably sterile, and it is to be feared that it will remain so, since the readers that could in principle be gained for a new, better, more complex science fiction could be won only from the ranks of the readers of mainstream literature, not from the ranks of the fans. I do not believe that it would be possible to read this hypothetical, nonexistent, and phenomenally good science fiction if you had not first read all the best and most complex works of world literature with joy (that is, without having been forced to read them). The revolutionary improvement of science fiction is therefore always endangered by the desertion of large masses of readers. And if neither authors nor readers wish such an event, the likelihood of a positive change in the field during the coming years must be considered as very small, as, indeed, almost zero. It would then be a phenomenon of the kind called in futurology "the changing of a complex trend," and such changes do not occur unless there are powerful factors arising out of the environment rather than out of the will and determination of a few individuals.

Postscript. Even the best science-fiction novels tend to show, in the development of the plot, variations in credibility greater than those to be found even in mediocre novels of other kinds. Although events impossible from an *objective-empirical* standpoint (such as a man springing over a wall seven meters high or a woman giving birth in two instead of nine months) do not appear in non-science-fiction novels, events equally impossible from a *speculative* standpoint (such as the totally unnecessary end-game in Disch's *Camp Concentration*) appear frequently in science fiction. To be sure, separating the unlikely from the likely (finding in the street a diamond the size of your fist as opposed to finding a lost hat) is much simpler when your standard of comparison is everyday things than it is when you are concerned with the consequences of fictive hypotheses. But though separating the likely from the unlikely in science fiction is difficult, it can be mastered. The art can be learned and taught. But since the lack of selective filters is accompanied by a corresponding lack in reader-evaluations, there are no pressures on authors for such an optimization of science fiction.

> *Translated from the German*
> *by Franz Rottensteiner*
> *and Bruce R. Gillespie,*
> *with R. D. Mullen and*
> *Darko Suvin*

SCIENCE FICTION: A HOPELESS CASE—WITH EXCEPTIONS

1

On reading *In Search of Wonder* by Damon Knight and *The Issue at Hand* by James Blish, a couple of questions, the answers to which can be found nowhere, came to my mind.[1]

---

1. This essay is a rewritten chapter ("Sociology of Science Fiction") from my *Fantastyka i futurologia* (Science Fiction and Futurology). I have polemically sharpened the original text in several instances, and added the later review of Dick's work, which is absent from the book. I confess that I made a blunder when I wrote this monograph, since then I knew only Dick's short stories and his *Do Androids Dream of Electric Sheep?* I believed that I could rely on reviews published in the fanzines of other novels by Dick, with the result that I considered him merely a "better van Vogt," which he is not. This mistake was due to

For example: in science fiction fandom rumor has it that science fiction is improving every year. If so, why does the average production, the lion's share of new productions, remain so bad?

Or: we do not lack definitions of this genre. However, we would look in vain for an explanation for the absence of a theoretical, generalizing critique of the genre, and a reason why the weak beginnings of such criticism can be found only in "fanzines," amateur magazines of very low circulation and small influence (if any at all) on the authors and publishers.

Furthermore: Blish and Knight agree that science-fiction readers cannot distinguish between a high-quality novel and a mediocre one. If they are right, how are readers selected to belong to the public who reads this literary genre, which intends to portray the (fantastically magnified) outstanding achievements of mankind?

The important question is: even if science fiction was born in the gutter, and lived on trash for years on end, why can't it get rid of the trash for good?

---

the state of science-fiction criticism. Every fifth or eighth book is praised as "the best work of science fiction in the whole world," its author is presented as "the greatest science-fiction author ever," great differences between works are minimized and annulled, so much so that in the end *Ubik* may be regarded as a novel that is just a little better than *Do Androids Dream of Electric Sheep?* Naturally, what I say does not justify my mistake, because no criticism is a substitute for reading the books concerned. However, my words still describe the circumstances responsible for causing my error, because it is a physical impossibility to read every science-fiction title, so that there must be a selection; as you can see, one cannot rely on science-fiction criticism to make this selection.

My essay tries to answer these questions. Therefore, it is a "Prolegomenon to Science-Fiction Ecology"—or an "Introduction to a Socioculturally Isolated Realm of Creative Work"—or a "Practical Guide for Survival in the Lower Realm of Literature." These pompous titles will be justified below. The books by Blish and Knight were of great assistance to me in writing, but I did not regard them as only collections of critiques, but more as ethnological protocols of several explorations into the exotic land of science fiction—i.e., as raw material to be subjected to a sociological analysis. For me, the facts collected by these authors were often more valuable than their opinions; that is to say, I arranged this material in a way not completely corresponding to the spirit of the sources.

2

I call science fiction a "collective phenomenon" of a sociocultural nature. It has the following parts: (a) the readers—on the one hand, the mute and passive majority of science-fiction consumers; on the other, the active amateur groups that constitute fandom proper; (b) the science-fiction producers—authors (some of them also critics) and publishers of magazines and books.

Science fiction is a "very special case" because it belongs to two distinct spheres of culture: the "Lower Realm," or Realm of Trivial Literature, and the "Upper Realm," or Realm of Mainstream Literature. To the Lower Realm belong the crime novel, the Western, the pseudo-historical novel, the sports novel, and the erotico-sentimental stories about certain locations, such as doctor-nurse romances, millionaire-and-the-playgirl stories, and so on. I'd like to spare the reader a detailed description of what I mean by mainstream. Perhaps it will suffice to quote the names of some of the authors who inhabit this Olympus: Moravia, Koestler, Joyce, Butor, Sartre, Grass, Mailer, Borges, Calvino, Malamud, Sarrault, Pinget, Greene.

It cannot be maintained with universal validity that these authors do not descend to the lower floors occasionally, for we know of crime novels by Graham Greene, "fantastic" novels by Orwell and Werfel, and Moravia's "fantasies." Some texts by Calvino are even considered science fiction. Therefore it should not be conceived that the difference between authors of the Upper and Lower Realms is that one of the first does not write fantasy or other literature related to science fiction, while one of the second does just this: the difference can be examined according to neither intrinsic type nor the artistic quality of a single work. To be a subject of the Lower or Upper Realm does not only and exclusively depend upon the product made by the author. There are much more complicated interrelationships of a sociocultural nature. I will talk about them a little later.

At this point I want only to propose a practical rule of procedure, which will predict with ninety-eight percent accuracy whether an author will be considered as an inhabitant of the Upper or the Lower floor. The rule is simple and can be stated as follows: if someone starts to write in the mainstream, and the public and critics get to know him by name, or even as a world celebrity (so that, on hearing the name, they know that they are talking about a writer, not an athlete or actor, so his attempts at science fiction and/or fantasy are regarded as "excursions" or "side leaps," even if repeated) then that man lives on the Upper floor. For instance, the "entertainments" of Graham Greene express a private mood or tactic of his.

During H. G. Wells's working life, there was no such clear-cut border between these two realms of literature. They shaded into each other gradually and continually. At that time Wells was known simply as an English writer, and the readers who appreciated his prose often knew of both his ambitions—the realistic and the fantastic. Only much later did an Iron Curtain descend between these two kinds of literature so that the typical science-fiction fan often knows the works of science fiction written by Wells, but ignores the fact that Wells also wrote "normal" realistic prose (and highbrow connoisseurs value it highly today, and more so than his science fiction). This curtain, this concrete ceiling (to maintain the image of a two-story building), has grown little by little, and this ceiling, hermetically sealed, became an impenetrable barrier only during the twenties. We can recognize this by the fact that Capek's works are still classed with the literature of the Upper Realm, while Stapledon, who was writing about ten years later, is not accredited with being there. Therefore some authors do not earn their classifications exclusively on their merits. On the contrary, their works are subject to higher rules of taxonomy, rules that have developed in the course of history and know *no* exceptions.

If, in spite of all this, a classificatory exception is made, the judgment is given that the (literary) case under consideration is *not essentially* science fiction, but wholly "nor-

mal" literature, which the author intentionally camou-
flaged as science fiction. However, if we proceed by disre-
garding all these "extenuating circumstances," some novels by Dostoevsky become "crime novels," though in fact
they are not regarded as such. The experts say that the
plot of a crime novel served the author only as a means
to an end, and he definitely did not want to write a crime
novel. This is the same situation as is the case of a brothel
searched by the police. For simplicity's sake the nameless,
ordinary guests are regarded as customers of the prostitutes, but a prince or a politician defends his presence on
the pretext that he descended to these lowest floors of
social life because he longed for something exotic, because
his fancy took him on such an excursion. In short, such
people stay in the land of pestilence as extravagant intrud-
ers or even as curious scientists.

3

The status of trivial literature can be recognized by
several typical attributes.

First: its works are read only once, just like the cheapest
mass products, which are also intended for but a single
use. Most of them become obsolete in the same way as
mass products do. If crime novels were selected according
to their literary merits, it would be superfluous to keep
throwing new ones onto the market, because we could find
so many good ones among the multitude there already
that nobody could read the choicest of them during his
lifetime. Still, publishers keep on putting "brand-new"
crime novels onto the market even though there are quan-

tities of crime novels of undisputedly better quality that have sunk into oblivion. The same goes for refrigerators and cars: it is a well-known fact that today's models are not necessarily better, technologically, than those of yesteryear. But in order to keep going, the machinery of production must put new models on the market, and advertising exerts pressure on the consumers to make them believe that only the current year's models have the best quality. The dogma of continual change of models becomes a law of the market, although every specialist can distinguish clearly between fictitious obsolescence of the product and authentic technological obsolescence. Off and on there are real improvements in technological products. More often, change is dictated only by fashion, a dictatorship in the interest of profit by supplying new goods.

The entanglement of real progress and economic laws constitutes a picture of a situation quite similar to that which reigns in the market of trivial literature. On principle, publishing houses like Ace Books could put on the market science fiction from the first half of the century exclusively, in ever-renewed reprints, because the number of this kind of book has already increased to such an extent that nobody could read even the better ones among them, even if he devoted all his time to this genre. Printing new books, ninety-eight percent of which are miserable products, published for purely economic reasons, makes many older works fall into oblivion. They die in silence, because there is no place for them on a clogged market. The publishing houses provide no filter to bring about a positive selection, because to them the newest book is also the best, or at least they want the customer to believe this, the justification for the well-known total inflation of publishers' advertising. Each new title is praised as the best in the science-fiction genre. Each science-fiction writer is called the greatest master of science fiction after one or two of his books have been published. In the science-fiction book market, as well as in the whole market of trivial literature, we can perceive the omnipotence of economic laws. The literary market, moreover, has in common with the whole market the typical phenomenon of inflation. When *all* books and writers are presented as "the best," then a devaluation, an inflation of *all* expressions of value is inevitable.

Compared with these carryings-on, with this escalation of advertising, the behavior of mainstream editors is quite shy and silent. Please compare the blurbs on the jackets of science-fiction books with those that serious publishers put on the jackets of a Saul Bellow or a William Faulkner. This remark seems to be banal, but it isn't. Although instant coffee or cigarettes of every brand are always praised as the best in the world (we never hear of anything advertised as "second best"), Michelangelo's frescoes and Tolstoy's *War and Peace* are not offered, with the same advertising expenditure, as the best works of art possible. The activities of the publishers of trivial literature make us recognize that this literature is subject to economic laws exclusively and to the exclusion of any other laws of behavior.

Second: I must remark that a reader of trivial literature behaves just like the consumer of mass products. Surely it does not occur to the producer of brooms, cars, or toilet paper to complain of the absence of correspondence, fraught with outpourings of the soul, that strikes a connection between him and the consumer of his products. Sometimes, however, these consumers happen to write angry letters to the producer to reproach him with the bad quality of the merchandise they bought. This bears a striking similarity to what James Blish describes in *The Issue at Hand,* and, indeed, this author, more than five million of whose books have been printed, said that he received only some dozens of letters from readers during his whole life as an author. These letters were exclusively fits of temper from people who were hurt in the soft spot of their opinions. It was the quality of the goods that offended them.

Third: the market of trivial literature knows only one index of quality: the measure of the sales figures of the books. When an "angry young critic" snubbed Asimov's *Nightfall and Other Stories* as old hat, Asimov put up the defense that his books, this year and for years previously, had sold excellently and that none of his books had been remaindered. Therefore he took literary merit for the relation of supply and demand, as if he were unaware that there have been world-famous books that have never been printed in large quantities. If we use this yardstick, Dostoevsky is no match for Agatha Christie. There are many fans of science fiction who have never read a novel by Stapledon or Wells in their lives, and with an easy mind I can assert that the silent majority of readers does not even know Stapledon by name. Blish and Knight agree that the public cannot distinguish a good novel from an abominable novel; and this is correct, with the proviso that they are talking about only the readers of the Lower Realm. If this generalization were valid for all readers at all times, we should have to consider the phenomenon of cultural selection in the history of literature as a miracle. For if all or almost all readers are passive and stupid beings, then who was able to collect Cervantes and Homer into the treasure troves of our culture?

Fourth: there are crass and embarrassing differences between the relations that link the authors of Upper and Lower Realms with publishers. In the Upper Realm it is the author who alone determines the title, length, form, and style of his work, and his right to do so is guaranteed unequivocally by the letter of his contract. In the Lower Realm, the publishers appropriate these rights. We can recognize from paragraphs of the printed contracts of large science-fiction publishing houses like Ace Books that it is the publishers who can, at their own discretion, change titles, length, and even the text of a book without express permission of its author, just as fancy dictates. Naturally, the editors of the Upper Realm also make encroachments. In practice these actions are quite different; they occur before the author signs the contract—i.e., first the editors propose to the author what they want changed, and only after he has agreed is the contract made, and not one syllable says that the original manuscript must be revised. The difference is because in the Upper Realm literary texts are considered in their integrity untouchable and taboo because they are almost sacred art objects. This is an old custom, in the spirit of the historical tradition of Western culture, though the practice of publishing, even in the Upper Realm, is not always so pious and fair as we are told. However, this difference between the two realms is of great importance.

In the Upper Realm one always strives at least to keep alive the appearance of intact virtue, in the same way as in high society women do not permit themselves to be called "prostitutes" although they indulge in open promiscuity. The "ladies" of the underworld do not have such pretensions, and it is no closely guarded secret that one can buy their favors at the appropriate price. Sad to relate, the authors of science fiction are quite similar in behavior to those "ladies," and they do not feel the disgrace of making transactions, either, as part of which they willingly hand over their works to the publishers, who are allowed to revise the texts at will. Thus James Blish[2] tells

---

2. It is quite difficult to shake off either a bad or a good tradition, once it is established. In *The Issue at Hand,* James Blish complains that English criticism surpasses American, and that this difference of level can be seen also on another plane—according to Blish, English publishers treat science-fiction authors with a consideration scarcely to be found in United States. His words date from the fifties. From what I know of the state of things today, this difference has decreased insofar as American criticism has improved, insignificantly, and English publishers have become a bit less considerate.

However, these particular differences should not make us wonder. American science fiction descends from the pulps; English science fiction had as its father, not Hugo Gernsback, about whom nobody outside of U.S. science fiction knows a thing, but H. G. Wells. What else? American science fiction worked itself up from the gutter of literature (though it could not fly into the sky); English science fiction has Americanized itself partly for commercial reasons, and partly has stepped into Wells's shoes, something that should not be taken as praiseworthy. The "classical" successor to Wells, John Wyndham, worked like a huckster, seeking to supplement the work of the master and teacher by filling what was, in his eyes, a gap. But even as anyone who paints like Van Gogh today cannot become a Van Gogh, so Wyndham did not add anything

us that his *A Case of Conscience* is only the length it is because his publisher at the time, owing to certain technical circumstances, could not produce a work of greater length! Just imagine if we read in the memoirs of Hermann Hesse that his *Steppenwolf* was only so long because his publishers. . . . Such a disclosure would cause a shout of wrath from literary circles, but Blish's words do not affect either him or any other author or critic because in the Lower Realm the station of a slave is taken for granted. Publishers are within their rights when changing the title, length, and style of science-fiction books as these encroachments are determined by economic considerations: they act like people who must find a purchaser for their goods, and they have a firm conviction that they work hand in glove with the author, like project leaders and advertising managers for Ford. Naturally, nobody thinks it strange that the project leader for a new model does not have the right to think up a name for it.

---

major to Wells's work. Wells worked according to the known principle of escalation, so that in *The War of the Worlds,* earth is attacked *only* by the Martians; but in Wyndham's *The Day of the Triffids,* the author does not think it sufficient to let all mankind go blind— he foists poisonous plants upon it; but because those plants do not seem dangerous enough, he adds the gift of active motion as spice.

After all, there are two distinct traditions in science fiction: the English, with the better manners and customs of the Upper Realm, and the American, which has lived from its beginnings in the slums of the Lower Realm, this slave market, which has no overabundance of courtly manners. Also, the language of English science fiction has always been more cultivated.

4

Seen in isolation, a number of the traits of trivial litera-ture, as described above, are quite unimportant. However, when added up, they form an ordered structure of the environment in which science fiction is born and gains a scanty living. These traits are clues, pointing out how in different ways the status of a work of literature is deter-mined; it depends upon whether it is born in the Upper or Lower Realm.

Thus science-fiction works belong to the Lower Realm —to trivial literature. Thus sociocultural analysis finally solves the problem. Thus words said about it are wasted; the trial can be closed with a sigh of relief.

But this is not so. Without a doubt there is a difference between science fiction and all the neighboring, often closely related, types of trivial literature. It is a whore, but a quite bashful one at that; moreover, a whore with an angel face. It prostitutes itself, but, like Dostoevsky's Sonya Marmeladova, with discomfort, disgust, and con-trary to its dreams and hopes.

True, science fiction is often a liar. It wants to be taken for something else, something different from what is really is. It lives in perpetual self-deception. It repeats its at-tempts to disguise itself. Has it got the shadow of a right to do so?

Many famous science-fiction authors are trying to pass for something better than their fellow writers—the au-thors of such trivial literature as crime novels or Westerns. These pretensions are often spoken out loud. Moreover, in the prefaces to their books, embarrassing praise is given to the authors by the authors themselves. For instance, Heinlein often emphasized that science fiction (that is, his own science fiction) was not only equal to, but also far better than mainstream literature, because writing science fiction is more difficult. Such pretensions cannot be found in the rest of the field of trivial literature.

This does not mean that there is no standard of quality for crime novels. Here, too, we distinguish bad, boring novels and original, fascinating ones. We *can* speak of a first-rate crime novel—but it does not occur to anybody to consider such a hit as equal to the masterpieces of literature. In its own class, in the Lower Realm, it may be a real diamond. When in fact a book does cross the borders of the genre, it is no longer called a crime novel, just as with a novel by Dostoevsky.

The best science-fiction novels want to smuggle themselves into the Upper Realm—but in 99.9 percent of cases, they do not succeed. The best authors behave like schizophrenics; they want to—and at the same time they do not want to—belong to the Realm of Science Fiction. They care a lot about the prizes given by the science-fiction ghetto. At the same time, they want their books to be published by those publishing houses that do not publish science fiction (so that one cannot see from the book jackets that their books are science-fiction books). On the other hand, they feel tied to fandom, write for fanzines, answer the questions of their interviewers, and take part in science-fiction conventions. On the other hand, publicly, they try to stress that they "do not really" write science fiction; they would write "better and more intellectual books" if only they did not have to bear so much pressure from the publishers and science-fiction magazines; they are thinking of moving into mainstream literature (Aldiss, Ballard, and several others).

Do they have any objective reasons for surrendering to frustration and feelings of oppression in the science-fiction ghetto? Crime novels are another, an open-and-shut, case. Naturally, a crime novel reports on murders, detectives, corpses, and trials; Westerns, on stalwart cowboys and insidious Indians. However, if we may believe its claims, a science-fiction book belongs at the top of world literature! For it reports on mankind's destiny, on the meaning of life in the cosmos, on the rise and fall of thousand-year-old civilizations: it brings forth a deluge of answers for the key questions of every reasoning being.

There is only one snag: in ninety-nine cases out of a hundred it fulfills its task with stupidity. It always promises too much, and it almost never keeps its word.

For this reason, science fiction is such a remarkable phenomenon. It comes from a whorehouse but it wants to break into the palace where the most sublime thoughts of human history are stored. From the time it was born, science fiction has been raised by narrow-minded slaveholders. Thomas Mann was allowed to work on one novel for fourteen years; John Brunner complains that there was a time when he had to write eight novels a year in order to stay alive comfortably. From shame, science fiction tries to keep some sides of this situation a well-guarded secret. (Often we hear from science-fiction authors how much freedom they enjoy in their work.)

Science fiction is subject to the rigid economic laws of supply and demand. It has not completely adapted itself to the "editor's milieu," meaning that there are recipes on how to write a science-fiction work that appeals to a certain editor and gains his appreciation (for instance, the late John W. Campbell, Jr., was an authoritative man who published only a certain easily definable kind of science fiction, and some authors knew how to foresee his demands). In Geis's *Science Fiction Review,* Perry A. Chapdelaine gives us a detailed account of how he was carefully briefed by well-known science-fiction authors when he wrote his first novel. Special care was taken to include those qualities that maximize sales; no mention was made of the immanent quality of the work itself. Often the same is the case in the Upper Realm—but only for beginners. However, science-fiction authors remain minors in the eyes of their publishers—all their lives. Such circumstances breed frustration and compensatory behavior. Indeed, the same sort of thing abounds in the science-fiction ghetto. All these compensatory phenomena, taken together, clearly have the character of mimicry.

(a) In the science-fiction ghetto there is no lack of makeshift and ersatz institutions which exist side by side with those of the Upper Realm. The Upper Realm has the Nobel Prize and other world-famous literary awards. The science-fiction ghetto has the Hugo and Nebula awards; and American science fiction poses (still) as "world" science fiction, as can be seen from anthology titles such as *The World's Best S/F.*

(b) The Upper Realm has academic and other highbrow literary journals, containing theoretical and hermeneutical articles. Science fiction also has its highbrow fanzines (*Riverside Quarterly* from Canada, *Science Fiction Com-*

mentary from Australia, and *Quarber Merkur* from Austria). These are parallel, although not analogous phenomena. The highbrow periodicals of the Upper Realm command real authority in cultural life. The most famous critics and theoreticians of the mainstream are all known to the cognoscenti and to almost all intelligent readers, at least by name (e.g., Sartre; Leslie Fiedler). Yet the names of the best science-fiction critics are not known to one soul outside the inner circle of fandom, and the silent majority of science-fiction readers does not know of the existence of the highbrow magazines. Even if they did know of them, they would not care for the evaluations of the cognoscenti—i.e., they are not influenced by these fanzines when choosing the new science-fiction books they are going to buy.

The *structure* of the flow of information is quite different in the Upper Realm than in the Lower Realm. In the Upper Realm the highbrow periodicals form the peak of a pyramid whose base is mass culture. The popular critics of the dailies need not agree with the judgments of the initiated highbrow experts, but if one of them opposes a man like Sartre, he knows quite well that he is fighting a world-wide authority. Nothing of this sort in science fiction. Its pyramid is hidden deep in the fan underground, the best fanzines have only insignificant circulations, and they cannot count on financial help from social or cultural institutions. (There are rare exceptions, such as *New Worlds,* which at one time received essential aid from certain British cultural institutions, but this is no longer the case in the United States.)

(c) Science-fiction conventions are intended to form a

kind of match for the meetings of the PEN Club and other similar gatherings. This also involves mimicry, because PEN meetings do not have in the slightest the character of a party that is so characteristic of science-fiction conventions. At conventions, theoretical reflections are nothing but seasoning; at PEN meetings, however, as well as at similar conferences of professional writers, they are the main course.

I must stress that no esoteric highbrow magazine of the Upper Realm has any direct influence on the policies of publishers. These magazines possess only a purely moral authority, founded on tradition. They do not try to wage open warfare upon the typical phenomena of mass culture today (e.g., normally they hide all data about one-day best sellers) and their activity becomes visible only in the long run, as all of the institutions in the structure feed the slow process of the Upper Realm. They should be the (often quite powerless) conscience and memory of world culture, its highest tribunal, which is at the same time an unbiased witness and judge. Often this tribunal loses a single skirmish but wins the great, epic wars—just the way Great Britain did. It cannot give a guarantee of today's fame to a great, misjudged poet, but it provides a memory, helping the next generation sometimes to dig up treasures that are almost lost. In short: these tribunals are not subject to the economic rules of the market, and because of this they are able to defend the cultural heritage against the chaotic onslaught of mass culture.

Nothing like that can be seen in the Lower Realm. Science fiction has no independent periodicals that supervise critically the whole production and form a similar fraction of the bulk of publications in the field, as in the case in the Upper Realm (measured by the yardstick of the circulation of books and especially of literary periodicals). The evidence of the best and best-known science-fiction authors is suppressed when it is contrary to the interests of the publishers—a fact that Knight reports on. The highbrow fanzines are known exclusively to a very small circle of initiated readers, and their influence on publishers' policies is nil. These amateur magazines often publish analyses and reflections that are equal in quality to the best of what is published in the Upper Realm. But this does not change the fact that no one listens to the voices of the critics. This important fact shows clearly that it is not the immanent quality of a statement that determines its scope of action, but this radius is contingent on the broader structure of the whole network of information with which the medium that published this statement is connected.[3] It is a typical

---

3. This does not mean that the radius of effective action of a statement varies directly with the range of a medium—i.e., in our case, that this radius grows in proportion to the increase of circulation of the periodical in which this statement is printed. In regard to circulation, many highbrow literary periodicals are no better off than the high-circulation fanzines, and the literary and theoretical publications of university faculties sometimes have tiny circulations, as low as three hundred or four hundred copies. What I am saying is that the degree of attention paid by the public to a "message" (a normative judgment) is determined by quite different factors from those of circulation. So, in some countries, an extreme degree of public opinion is paid to several "underground" papers, though these pamphlets look shabby and are circulated in very tiny editions. The authority, the weight, of such statements belongs to the imponderabilities of civilization; the public must be aware in advance that somebody important has something to say;

science-fiction custom that critiques are not produced independently, but are written by either the authors or the editors of anthologies, who evaluate each other's works. This state of affairs only helps to cloud the line of demarcation between apologetics (a public-relations affair) and objective criticism.

Taken as a whole, science-fiction institutions (cons, fanzines, and awards) appear similar to those of the Upper Realm, but dissimilar as regards the function of furthering social values and selections. In the Upper Realm, as time goes by, the worst and the best literary works drift apart from each other; in science fiction however, the forces that are the result of economic laws of the marketplace, an absence of independent criticism, and a lack of cultural assistance are all directed toward the opposite tendency. They put trash next to valuable books; they impede any experiments in literary creation, choke independent, demanding, probing criticism, and they assist publishers in

---

but the "inherently wise," or even the "eggheads," do not possess such authority and attraction in their own right. The channels that serve to disseminate information are not built by technical and material means (such as the number of copies of a periodical distributed). Instead, these copies find their own way and have their maximum effect only if they flow into a broader structure that strengthens the message. This is the case for the highbrow periodicals because they live at the peak of the cultural pyramid. It is an extremely important phenomenon, which has been almost neglected. In many circles of fandom, people believe that one could wake the "silent majority" of the public from its slumber if only one could bomb the public incessantly with beautifully made publications with mass circulations. Most probably the public would throw these fine pamphlets into the nearest wastepaper basket because this bombardment of mass-produced science fiction would still lack the necessary influence. Authority and influence are not acquired easily.

camouflaging as true criticism the advertising that boosts the sales of their products.

Furthermore, the chain of publishers who specialize in science fiction—and the silent majority of mute, passive readers—forms an environment to which even the most gifted science-fiction authors must adapt themselves eventually. The authors are initiated early into the rules of the game, and they must either obey or take immense risks. Suppose an ingenious, even inspired author enters the realm of science fiction. This man must adapt rapidly and without scruples to the simple truth that it is impossible for him to be valued and esteemed according to his extraordinary achievements. The silent majority of the readership will devour his valuable books in just the same way, at best, as they are used to absorbing the worst nonsense of mass production. Taking into account just the economic barometer of the market, the publishers will treat him in the same way as they treat his colleagues—i.e., as authors who willingly allow the titles, lengths, and structures of their books to be changed in advance according to the wishes of their masters. This author will watch helplessly the embarrassing sight of his books submerging in an ocean of trash, for the stigma of science fiction links them irrevocably to this sea. Surely Sturgeon is right in maintaining that ninety-nine percent of all books in every genre are trash, but the fact remains that in the Upper Realm of culture there are forces that never cease furthering positive selection. In the Lower Realm, the best books are placed beside the worst and most stupid, and submerged by them under the pressure of the objective situation.

Thus, science-fiction institutions only seem to be the equal of the institutions of the Upper Realm. In fact, we see before us a superficial mimicry. Science fiction merely apes and simulates the Olympian quality of literature, without reproducing the same performance capability. No famous author from the Upper Realm concerns himself with disqualifying trashy literature or defending himself against the attacks of graphomaniacs. For a while, the Knights and Blishes tried to do this, but in the end their aggressiveness had to give way to a moderated, more passive attitude. To some extent these intelligent men are conscious of their own defeat. They feel that this behavior, typical of science fiction, merely apes grown-up literature. They can see how grotesque such goings-on must look to an outside observer. The unauthenticated (because not earnest) quality of fandom, with its letters, parties, and friendly exchange of opinions, is for the authors only a weak substitute, an asylum where they can play the part of the great writer by confessing in fanzines with circulations of two hundred or less the secret of their creative writing and their deep psychological secrets.

We could consider these phenomena as insignificant and pay no attention to them, because in the end the ways in which the literati compensate their inferiority complexes, their feelings of frustration, and their *Wille zur Macht* are not necessarily those aspects of literature that flourish in the Upper Realm. However, in the Lower Realm these are symptoms of the chronic illness that impedes so embarrassingly the growth of the science-fiction genre. Thus the only way to better the prevailing situation is to make an outspoken diagnosis. We could support this conclusion with hundreds of examples. In an article by a contemporary science-fiction critic, the names of authors, including Farmer, Joyce, Sturgeon, and Kafka, are listed indiscriminately. But mainstream critics never reciprocate this striving for equal status. In today's science-fiction anthologies we find, apart from science-fiction authors, such writers as Grass, Calvino, Ionesco, and Michaux, but the Upper Realm does not offer any just return. The inhabitants of the Upper Realm are invited to the Lower; they accept these invitations, but there is no return service. The inhabitants of the Upper Realm treat those of the Lower Realm properly, just as the gentry treat the rabble properly. A lady may enter a honky-tonk, but the "ladies" who reside there permanently are not allowed into a respectable house.

5

We shall now show how the work of a gifted science-fiction writer grows in the science-fiction environment and how it is accepted there. (The fate of the untalented does not concern us—but we will report on it, too, if only marginally, as it turns out in quite a characteristic way in the Lower Realm.)

The substance that fills the entire milieu of science fiction, and upon which the work of its authors feeds, is kitsch. It is the last, degenerate form of myths. From them it inherited a rigid structure. In myth the story of Ulysses is the prestabilized structure of fate: in kitsch it becomes a cliché. Superman is a spoiled Hercules, the robot a golem, even as kitsch itself is the simplified, threadbare, prostituted, but original constellation of values central to a given culture. In our culture, kitsch is what was once holy and/or coveted, awe-inspiring, or horrible, but now prepared for instant use. Kitsch is the former temple that has been so thoroughly defiled by infidels for so long that even the memory of its ancient untouchability has been lost. When hitherto untouchable idols get the status of mass products, through mechanical reproduction, and become obtainable as everybody's objects of enjoyment, we observe how the originally sublime is degradingly transubstantiated into kitsch. The venerable paradigm is reworked in order to make it easily consumed and as simple as possible. And—quite important—kitsch does not present itself as such to its consumers; it believes in its own perfection and wants to be taken seriously. Even the psychic process that originally kept the mass of the uninitiated at a distance from the object of worship, because it was an obstacle that had to be overcome, comes wrapped up with the goods as an appetizer. Kitsch, free from all difficulties of consumption, is a product that has been prechewed for the consumer. In literature, kitsch results when all the complexity, multi-sidedness, and ambiguity of the authentic product is eliminated from the final product.

However, the people concerned (both authors and customers) have a splendid feeling of well-being if this final product retains the air of being an *objet d'art*, in full bloom, without restrictions. Kitsch is composed exclusively of ersatz products: of heroism, of need, misfortune, love, etc. In science fiction, kitsch is made from ersatz science and literature. From reading "inner circle" critiques and considering what science-fiction pro-spectuses have to offer, you would hardly believe that the authors who are reviewed display an abundant ignorance of the grammar, syntax, and style of their mother tongue; it is as if one suddenly hears that a team of athletes preparing for the Olympic Games cannot yet get up and stand.

In a stabilized culture, the sphere that kitsch might inhabit is quite small. In mass culture, it tends to overflow into neighboring genres; it has an aggressive and explosive pressure; it is a tumor that grows exuberantly, devouring that part of the body which is still intact. It is quite hard to justify morally a defense against its attacks, because the dilemma always arises as to which is the lesser evil: the trashy deformation of an art object, or its total absence from the circuit of a mass culture that cannot assimilate the real thing. Science fiction is a clinical case of a region occupied exclusively by trash, because in kitsch, the culturally and historically highest, most difficult, and most important objects are produced on the assembly line, in the most primitive forms, to be sold to the public at bargain prices.

Knowing no discretion and no reverence for things inconceivable by the human mind, piling universes upon universes without batting an eyelash, mixing up physics, metaphysics, and trite trash from misinterpreted philosophical systems without end, science fiction is the true embodiment of kitsch, because of the cheekiness of its total ignorance, which even denies the existence of a higher knowledge, toward which it finds no path, and denies it triumphantly and obstinately.

Even if there are subjects about which philosophers dare not even think, topics about which world-famous scholars can say scarcely anything at all, they can be bought for 75¢ to $1.25 at every newsstand for immediate inspection. Science fiction provides a pleasant substitute for the study of the handbooks of the greatest thinkers, cosmologists, astrophysicists, and philosophers who have ever lived—yes, it can even report on what scientists born a thousand years from now will know. I am not ridiculing this maximum offer; I can only repeat what you read in the science-fiction advertisements. If somebody ridicules somebody else, you could not tell from the earnestness of these statements; it is just another case when you can't take a single word seriously, for this is advertising, which is used to talk only about the best possible and previously nonexistent products. If all this is not meant to be taken seriously, then what is the real content of all their cipher language?

One of the most incredible secrets of science fiction (although one not too closely guarded) is the fact that ninety-nine percent of its authors do not know even the titles and authors of today's learned works, but still they want to top these scholars with their knowledge of the year 6000. If an author understands schoolteacher's physics, he is praised by Knight, quite in earnest, and presented as a model to authors who seem to have been forced to drop out of school after three years because of general mental weakness. The public does not seem to wait to find out about these interesting facts, probably because such news would annoy them. It is quite embarrassing to find out that for the least amount of money and mental effort, one has been convinced that one was initiated into the vastest secrets of the universe and existence.

6

The exception mentioned in the title of this essay is the work of Philip K. Dick. Because of the lack of a selection process to struggle against trash and promote real value, the works of Dick are sometimes compared with those of A. E. van Vogt.

The novels of both authors share the common characteristics that (1) they are composed of trashy parts and (2) they are full of contradictory elements. The contradictions include those of an external nature (as when the world depicted in a book runs counter to empirical scientific knowledge) and of an internal nature (as when during the course of a novel the action becomes self-denying—i.e., contradicts itself).

Such a diagnosis does not automatically invoke a subsequent condemnation. It is true that literary judgment is undemocratic, but nevertheless in the course of each critical trial it is also just. Yet it must be ascertained why the case under scrutiny allows a sacrifice of values. These works contain local nonsense and a local destruction of values (sense is always to be preferred to nonsense), but this local inroad might aid the construction of a higher sense of the totality. This point is connected with the general relativity of all values: even a murder may be justified in a civilization where it is considered a link in a chain of connections in which, according to prevailing belief, the lesser value, a man's life, is sacrificed to the greater, the godhead.

Judged prima facie, there are no relevant differences between the two cases under review. Both authors disregard empirical knowledge, logic, and causality, categories upon which our knowledge is founded. They seem to sacrifice these basic values to the momentary stage effect; therefore, they destroy the greater values in order to create a lesser one—something always culturally taboo.

However, our authors are writers of quite different ranks, when read thoughtfully. As Knight and Blish have proved, the phantasmagoric acrobatics of van Vogt do not add up to a meaningful whole. He does not solve the riddles posed, he does not draw conclusions from the things depicted early in his books, and he sketches only ephemeral ideas, piling them chaotically on one another. With all that, he does not hypnotize the wary reader, but only lulls him to sleep; this sleep comes from increasing boredom, not fascinating magnetism. The only problem posed by van Vogt's prose is its financial success, at the same time irritating and annoying an intelligent reader like Knight. Why is it possible that work the stupidity of which was amply and unequivocally demonstrated by Knight still enjoys such great popularity?

But no deep secret awaits discovery. The van Vogt fans do not care a jot about the Knight line of deduction. Most probably they do not know of it and do not want to, either. From van Vogt they get the whole cosmos, with its inhabitants, wars, and empires, excellently served up, because the plot can be seen without thinking at all, and they close their eyes to the knowledge that they are being fed with stupid lies. We will say no more on this topic.

Philip K. Dick seems to write in a vein similar to van Vogt's, although he does not, like van Vogt, violate grammar and syntax as well as physics. Dick, too, works with trash. Yet his novels are structured with more logic. He is accustomed to let action issue from a clearly and precisely built situation, and only later in the course of a novel does decay, perplexing the reader, begin to undermine initial order so that the end of the novel becomes a single knot of fantasies. Dreaming and waking are mixed, reality becomes indistinguishable from hallucination, and the intangible center of Dick's world dissolves into a series of quivering, mocking monstrosities so that in the end each novel of Dick's mainstream (for Dick has also written second-rate, insignificant works) destroys the order of things that he erected at the beginning. Even if Dick's worlds owe their explosion to a technology or a disease (or madness) of the space-time manifold, in ever-increasing speed they multiply their "pseudo-realities" so that (as in *The Three Stigmata of Palmer Eldritch*) the levels of hallucination and reality, which initially were separate from one another, become a space-time labyrinth. But Dick always moves among the typical trash of science fiction in the realm of androids, of the usual prophets ("precogs"), "psi," "esp"-fields, brain transplants, and hundreds of other similarly scurrilous products and phenomena.

Trash is present everywhere in Dick's books; from time to time, though, in some of his novels, he succeeds in executing a master stroke. I am convinced that he made this discovery unconsciously and unintentionally. He has invented an extremely refined tactic: he uses elements of trash (that is, those degenerate molecules that once had a sacramental, metaphysical value) so that he leads to a gradual *resurrection* of the long-extinct, metaphysical-exotic values. In a way, he makes trash battle against trash. He does not deny it, he does not throw it away, but he builds from it a ladder that leads straight into that horrible heaven, which, during this operation, ceases to be an "orthodox" heaven, but does not become an "orthodox" hell. The accumulating, mutually negating spheres of existence enforce the resurrection of a power that has been buried for eons. In short, Dick succeeds in changing a circus tent into a temple, and during this process the reader may experience catharsis. It is extremely difficult to grasp analytically the means that make it possible for him to do so.

On the contrary, it is easy to say that this catharsis justifies the sacrifice of values that shocks the reader at the beginning. I cannot devote this essay to the Dick Transubstantiation Method; therefore, I will make only a few remarks on his *tour d'adresse*.

The promise of "almightiness" is implicit in science fiction. This omnipotence has a bipolar nature—the omnipotence of the bad (as in the dystopia) and of the good (the utopia). In the course of its evolution science fiction has renounced the positive omnipotence, and for a long time it has occupied the opposite pole—that of maximum despair. Gradually it has made this pole its playground. Because the end of the world, the atomic Last Judgment, the epidemic provoked by technology, the freezing, drying up, crystallization, burning, sinking, the automation of the world, and so on no longer have any meaning in science fiction today. They lost their meaning because they underwent the typical inflation that changes eschatological horror into the pleasant creeps. Every self-respecting fan owns a science-fiction library of the agonies of mankind that equals the book collection of a chess amateur, since the end of the world should be as formally elegant as a well-thought-out gambit. I believe it is a very sad phenomenon to witness the indifferent workmanship with which such novels are produced. There are specialists who have slaughtered mankind in thirty different ways, but still search diligently and calmly for further methods of murder. Structurally this (end-of-the-world) science fiction has put itself on the same level as the crime novel, and culturally it acts out a nihilism that liquidates horror, according to the law of diminishing returns.[4] A space

---

4. This point of view may prompt some fans to ask the question why science-fiction writers should not be allowed to make an intellectual game out of the topic of mankind's doom, and why the science-fiction field should be forbidden that which is done with complete justification in the field of the crime novel? My answer is: Surely nothing in heaven or on earth prohibits us from doing so; in the same way as there are no "absolute" prohibitions to hinder us from playing with corpses or the genitalia of our fathers or from concentrating our whole love life on the goal of sleeping as fast as possible with as many women as possible in order to establish a record. We could do all these things as a matter of course, but surely nobody praises such programs as something to further social values: neither can we deny that these actions promise certain new liberties only annulling forever taboos that have stayed intact until today. As the English put it: you cannot have it both ways; you cannot respect a life, a topic, a feeling, and prostitute it at the same time. At the utmost you can falsify the real appearance and real meaning of a situation brought about by your own actions deliberately or unconsciously; but hiding one's head in the sand is fraught with well-known dangers. According to the whole historical tradition of our

occupied by trash is a vacuum in which lead and feathers fall at the same speed. It is indeed a great venture to coerce the resurrection of dead metaphysical values from such a novel.

It cannot be maintained that Dick has evaded all the traps set for him: he has more defeats than victories in his work, but the latter determine his rank as an author. His successes are due to his intuition. Average science-fiction authors form their hells of existence, their flaming grounds to head for, in social institutions, especially police-tyrannies-plus-brainwashing, as from Orwell's school, but Dick makes his out of ontological categories. The primary ontological elements—space and time—are Dick's instruments of torture, which he uses with great versatility. In his novels he constructs hypotheses that are prima facie wholly nonsensical (because of the contradictions they contain)—worlds that are at the same time determinist and indeterminist, worlds where past, present, and future "devour" each other, a world in which one can be dead and alive at the same time, and so on.

But in the first world even the "precogs" prove to be powerless to evade their own cruel end, which they foresaw themselves. Their wonderful gift only makes their

---

culture, truth has inherent value, whether pleasant or depressing. If crime novels follow their own schemata to falsify reality, it does not matter, since nobody looks into these novels for the highest revelations and initiations into the abysses of human nature. If science fiction adapts itself to the crime novel, it must stop claiming to be considered as something better than the crime novel. Its peculiar state of continual oscillation between the Upper and the Lower Realms of literature is a symptom of its repetitive attempts to have it both ways. But this is impossible without self-deception.

torture harder to bear. In the second world time becomes a Laocoon's snake that strangles its inhabitants. The third world embodies the saving of Chiang Tsi, who, upon waking, posed the famous question of whether he is Chiang Tsi who has just dreamed he was a butterfly or a butterfly who now dreams that he is Chiang Tsi. Dick writes about a technological realization of an ontological problem that has always occupied philosophers (i.e., the controversy between subjectivists and objectivists) so that it may be considered as an earnest problem of the (far) future, and not just a speculative question.

The common opinion that philosophical problems can never change directly into technological feats is an illusion caused by the relatively brief period of the technological era. In the year 1963 I discussed this problem in my *Summa Technologiae,* in the chapter entitled "Phantomatics." One possible way to build a synthetic reality is to "encapsulate" the consciousness by connecting the brain of the person in question to a computerlike apparatus in the same way as it is connected normally to real environments through the senses and nerves—i.e., with feedback. The most interesting puzzle is whether a "phantomatically imprisoned" man can divine the real state of things, whether he can distinguish the machine-simulated environment from the real one, by means of any one experiment. From either a logical or an empirical standpoint, it seems that the person could not make a correct diagnosis if the program of the machine was insufficiently developed. In a civilization which has such phantomatic techniques, there may be much mind-napping. But also there may be many legal uses of such methods, so that a person could witness while awake as many happenings as could be programmed, and since in principle there are no obstacles, the phantomated person would realize the counterempirical (he could, subjectively, live through many metamorphoses of his body).

In Dick's book *Ubik,* we find a literary variant of a similar project. He deals with a biotechnological method that is complicated by the fact that it allows dying people to remain in a specific state between life and death—i.e., "half-life." Dick develops a quite horrible game, so that it is not clear at the end which of the main characters lie in half-life and which live in normal reality. The action zigzags. With different ideas of what the reader is led to believe to be true. Also there are such macabre effects as the dissolution of earth and jumping back in time. You can find similar things in science fiction, but this masterly, gripping guidance of the play, in particular the behavior of all the characters, is psychologically depicted without fault. The border that separates the adventure novel from mainstream literature is transgressed in *Ubik,* something I want to prove later in this essay.

Now I want to review the "message" that several of Dick's novels communicate to us in an unequivocal way, embedded in the action. He seems to want to prove an equation, in the form of "We exist, therefore we are damned," and this equation is supposed to be valid for all worlds, even for impossible ones. His novels are the results of pessimistic ontological speculations about how the face of men would change if total revolutions in the basic categories of existence occurred (e.g., revolutions in the space-time system, in the relationship between dreaming and waking, etc). The result is the same, since insofar as these changes are induced by biotechnologies or drugs (as in *Palmer Eldritch*) they can only worsen the fatality of earthly existence. The greater a technological innovation is, says Dick, the more horrible its consequences.

In his first "major" novel, *Solar Lottery*, Dick has not yet tried to destroy the fundamentals of existence completely. He "shyly" introduces a new sociotechnology in which all men are supposed to have an equal chance to gain political power, because the allocation of power depends upon a comprehensive lottery. As can be expected, the result is a new kind of misery and inequality. Thus Dick has good reasons to sacrifice logic and causality; he shows that even the variants of existence that violate causality and logic are inherent in the invariance of texture and doom. One could call Dick an inverted apologist for "progress," because he connects unlimited progress in the field of the instrumentally realizable with bottomless pessimism in the field of human consequences of such progress in civilization. His novels are pieces of fantastic belles-lettres, but his underlying philosophy of life is not fantasy. Dick seems to foresee a future in which abstract and high-brow dilemmas of academic philosophy will descend into the street so that every pedestrian will be forced to solve for himself such contradictory problems as "objectivity" or "subjectivity," because his life will depend upon the result. With all his "precogs," "cold-packs," and "Penfields," he tells us, "And if you could achieve the impossible, it would not alleviate your misery one bit."

Dick's main characters are engaged in a battle not only for their lives, but also to save the basic categories of existence. They are doomed to failure in advance. Some exhibit the patience of Job, who gazed quietly into the face of what was coming, since everything that can happen to a man had already happened to him. Others are valiant wrestlers, striving after power, while still others are small and petty people, officials, and employees. Dick mans all his misleading worlds with contemporary Americans. Probably this is the reason why they seem so living and authentic—because there is a feedback between them and the world surrounding them. The authenticity of these people corroborates the fantastic background, and, vice versa, the background makes the normal people seem especially noteworthy and true to life. Dick's main characters do not become greater during the apocalyptically terrifying action of his novels; they only seem greater—or more human—because the world around them gets ever more inhuman (that is, more incomprehensible to the mind of man).

There are moments when they have a tragic effect. In the Greek sense tragedy is inescapable defeat, with several ways of being defeated. Some of these ways, if a man chooses one of them, give the opportunity symbolically to save an inestimable value. For one of Dick's heroes, the love of a woman or a similar human feeling is the kind of value that is worth saving, a value to be guarded even if the world goes to pieces. They are the last islands of spiritual sanity in a world gone mad, a world that heaps on them objects used in ways other than originally intended and that thus become instruments of torture and objects that spring from the sphere of the most trivial consumer goods and behave like things obsessed (e.g., a tape recorder or a spray can). Dick's main characters engage in conferences with monsters, which, however, are not little BEMs ("bug-eyed monsters," the embodiment of trash), because an aura of grotesque and dramatic worth clings to them, and they have the dignity of misshapen, tortured creatures. With the example of such monsters— one of which is Palmer Eldritch—we can see how Dick vanquishes truth; in the shape of a mutilation, he makes simple the macabre and the primitive by giving them a trace of fragile humanity.

In *Ubik,* the twitching world reminds us of the "will" of Schopenhauer, will gone mad; spurned into everlasting time explosion and implosion, devouring itself. As an aside, measured by the yardstick of Dick's black pessimism, Schopenhauer's philosophy of life seems to be real *joie de vivre.* Dick sees our world as the best of the worst, and there are no other worlds. According to him, we are everywhere damned, even where we cannot go. Dick once said that he does not consider himself a limitless pessimist. Possibly, though conscious of reason in the cosmos, he does not draw the nihilistic conclusion because he does not ascribe an exclusively negative value to the agony of man. But this is my private speculation.[5]

Dick's planets, galaxies, men, children, monsters, elevators, and refrigerators are all symbols of a language that, mix it as you please, always crystallizes into the same form of a *mene tekel.*

With that I don't want to say that Dick's novels—even

---

5. This applies only to the novels by Dick that I know: *Solar Lottery, The Three Stigmata of Palmer Eldritch, Our Friends from Frolix 8, Now Wait for Last Year, Do Androids Dream of Electric Sheep?, Ubik,* and *Galactic Pot-Healer.* In addition, I have read several of Dick's short stories, mainly in science-fiction magazines.

his best, like *Ubik*—are faultless masterpieces. The surfaces of his books seem quite coarse and raw to me, connected with an omnipresence of trash. I like what he has to say in one chapter more than what a page shows, and that is why his work forces me into fast reading. Upon looking his details in the face, one beholds several inconsistencies, as in looking at an Impressionist's painting from too close. Dick cannot tame trash; rather, he lets loose a pandemonium and lets it calm down on its way. His metaphysics often slips in the direction of cheap circus tricks. His prose is threatened by uncontrolled outgrowths, especially when it boils over into long series of fantastic freaks, and therefore loses all its function of message. Also, he is prone to penetrate so deeply into the monstrosities he has invented that an inversion of effect results: that which was intended to strike us with horror appears merely ridiculous, or even stupid.

With that I'll stop this immanent review of Dick's work and pass over to its sociological aspect. The science-fiction environment is unable to separate and make distinct the types of works that are being born into it. This environment is incapable of distinguishing clearly between the work of Dick, which is artistically bunched together into sense, from that of van Vogt, which collapses into nonsense. On a higher plane a title like *The World of Null-A* belongs to Dick, not to van Vogt, although it was the latter who actually wrote it; but only with Dick can we talk about a "non-Aristotelian" logic, whereas this title is merely tacked onto van Vogt's book without any justification. In its actions the science-fiction environment is by no means chaotic; obeying its own laws and regulations, it extols the stupid and denigrates the valuable until both meet "halfway"—on the level of insignificant trifles. In science fiction, Dick has not been honored according to his merits. Some people acknowledged the entertainment values of his novels, and one of the best living science-fiction critics, Damon Knight, also spoke about Dick's distorted pictures of contemporary reality (in *In Search of Wonder*) when he reviewed *Solar Lottery* and some other early books by Dick.

But that was all the praise this author came to hear. Nobody saw that his "unchecked growth" is quite strikingly similar in content and form to what goes on in the Upper Realm. Judged according to the problems he deals with, Dick's novels belong to that stream of literature that explores the no man's land between being and nothing— in the double sense.

(a) We can count Dick's novels as part of the prose that is today called "Literature of Ideas" or "Literature of Possibilities." This type of experimental prose tries to probe the neglected, latent, untouched, as-yet-unrealized potentialities of human existence, mainly in the psychological sphere. Probably one can find fountains of such prose in, among others, the works of Musil *(Der Mann Ohne Eigenschaften*—The Man Without Qualities) in which the outer world, randomly manifesting itself, affixes to the individual, so that he remains a soul "without qualities." In such books as his *Le Voyeur*, Robbe-Grillet tries other tactics; this prose seems to fit the motto *Quod autem potest esse totaliter aliter*—"that which, however, can be something wholly different" (which in Poland is represented by J. Andrzejewski in his *Miazga*, a work that is

written partly in the future subjunctive mood and therefore describes what could possibly happen, and not what
has unconditionally happened), which has its parallels
with Dick's work. Robbe-Grillet proceeds from the typical science-fiction blueprint of "parallel worlds," but
whereas most science-fiction writers flatten this motif into
unbearable trash, running over it like a steamroller, Dick
knows how to raise the problems that grow from this
inspiration to a fitting level of complexity. Therefore he is
an original representative of the Literature of Ideas in
science fiction—a wide field, but one with which I cannot
deal here exhaustively.

(b) In connection with Dick, we can think of authors
like Beckett, because of the "unhealthy curiosity" that
both have for death, or, more exactly, for the flow of life
as it approaches its end. Beckett "is content" with natural
processes that will devour man from the inside, slowly and
continually (as when growing old, or becoming a cripple).
Dick devotes himself to grander speculations, in the true
spirit of the genre he is working in.

We could say many interesting things about his "theory" of half-life (not as a sensible empirical hypothesis, but
as a variety of fantastic-ontological speculation) but, once
again, I cannot dig too deep into an exegesis of a desacral-
ized eschatology.

We draw these two parallels to show how an area of
creation, closed into a ghetto, suffers from the situation of
its own isolation. For such parallel courses of evolution
are not accidental coincidences. It is the spirit of time that
mirrors itself in them, but science fiction knows only
short-lived fashions.

The peculiarity of Dick's work throws a glaring light upon relationships within the science-fiction milieu. All science-fiction works have to give the reader the impression of being easy to read, as has all fiction. Science-fiction works before which two hundred Nobel Prize winners in the department of physics kneel down are worthless for the science-fiction market if, in fact, the precondition of being able to evaluate a work of science fiction is a minimum of knowledge. Therefore it is best for science-fiction books not to contain any deep meaning—either physical or metaphysical. But if the author smuggles any sense into his work, it must not stir the phlegmatic and indolent reader, or else this invaluable man will stop reading because of a headache.[6] The deeper meaning is admitted only if it is "harmless," if we can neglect it entirely while reading. The following anecdote may explain this problem: If many colored flags are put upon the masts of a ship in the harbor, a child on the shore will think that this is a merry

6. Each society is stratified according to its own pattern. In each society there are powers of selection with local effects to attract and repel individuals. Among others, such mass processes give rise to different readerships for widely differing varieties of literature. If one compared the intelligence and level of education of the average science-fiction reader in the United States and in the Soviet Union, one would draw the conclusion that the Russians know more about literature and are more intelligent than the Americans. However, this would be a fallacy; the selection processes of science-fiction readership in Russia and in the United States have taken different courses, because of the different traditions that prevail in the two countries in regard to the broader questions of the duties and psychosociological status that literature, as a whole, must play in society. Certainly the United States has the same percentage of bright boys and girls as Russia has but intelligent readers there approach science fiction far less often than they do in Russia.

game and perhaps will have a lot of fun watching, although at the same time an adult will recognize the flags as a language of signals, and know that it stands for a report on a plague that has broken out on board the ship. The science-fiction readership equals the child, not the adult, in the story.

Their trashy surface helps Dick's novels to survive in the milieu of science fiction. I do not maintain that Dick is a Machiavelli of science fiction who, under the cover of science-fiction trash, intentionally carries out a perfidiously thought-out camouflage in order to deceive his readers (i.e., in giving them gold disguised as iron trinkets).

Rather, I believe that Dick works intuitively, without knowing himself that he plays hide-and-seek with his readers. Please note the difference between an artist and an artisan: the artist grows in his environment, deriving from it the elements that serve him as a medium of expression—of those differences of tensions to which his personality is subject. The artisan is a producer of things for which there is a demand and which he has learned to produce—after the models that enjoy the highest popularity. Ninety-eight percent of science fiction is a craft, and its authors are day laborers who must obey to demand payment. Almost any artist can become an artisan when he strangles his inner voice—or he has no such voice at all.

For a long time Philip K. Dick has been only an artisan, and a skillful one, too, since he knew how to produce the things that were bought immediately. Gradually he began —and I must continue to speak in metaphors—to listen to his inner voice, and, though he still made use of those elements that science fiction put at his disposal, he began to put together patterns of his own.

But this is not an infallible explanation. As is always the case, it arises from a kind of cross-breeding between what is in the books I read and what I can do with this material as a reader. Therefore I can imagine other explanations for Dick's novels, explanations that differ from mine, though naturally the role of such an explanation cannot be played by just any idea. There is no doubt about the fact that with trashy elements Dick tries to express a metaphysics of an extremely "black" nature, mirroring authentically the state of his mind. A logical, one-hundred-percent un-equivocal reconstruction of the deep semantic structures of a complex work is impossible because there are no discursive series of phrases to which a work of art may be reduced without leaving something remaining.

Thus it must be; for if it were otherwise, this essay would be entirely superfluous. Why should I talk in so complicated and obscure a manner about a theme, if this theme may be put into clear and simple words? That which you can say briefly and intelligibly you need not describe with long and unintelligible words. For this rea-son, every authentic work of art has its depths, and the possibility that such a work of art carries a message about existence for subsequent generations of readers, although in society, in civilization, and in life there is endless change, bears witness that the transitory things that do not disappear in a masterpiece are buried in its semantic varia-bility. Out of the glaring clichés of trash, behind which yawns a horrible vacuum for every science-fiction artisan, Dick makes for himself a set of messages—i.e., a language —just like somebody who puts together from separate colored flags a language of signals according to his own judgment. Science-fiction criticism could help Dick to collect the colored flags, but not to put together sensible entireties from this crude material, because in practice it denies the existence of semantic depth.

Those science-fiction readers who are keenest of hearing feel that Dick is "different"; however, they are unable to articulate this impression clearly.

Dick has adapted to the science-fiction milieu—with positive as well as negative effects. He invented a method to express, with the aid of trash, that which transcends all trash. But he was unable to withstand to the end the contaminating influence of this quite poisonous material.

The most striking lack is the lack of penetrating, detailed, and objective criticism. The critical books by Blish and Knight are an exception to this rule; the book by Lundwall (*Science Fiction: What It's All About,* 1970) is not a piece of criticism or a monograph, but is merely a traveler's guide to the provinces of science fiction. The innocent sin of Blish and Knight is that they only and simply reviewed current science-fiction production, paying attention to all the authors. In their length and detail, the negative, destructive critiques written by Knight are totally superfluous, because it is impossible to help authors who are nitwits, and, as I said before, the public does not give a damn about such disqualifications.

Literature has no equality of rights: the day laborers must be dealt with in one sentence, if not with scornful silence, and a maximum of patience and attention is due to the promising author. But science fiction has different customs. I am no enthusiast; I do not believe that shrewd critiques would make author Dick into a Thomas Mann of science fiction. And yet it is a pity that there has been no critical selection among his works (although this state of affairs is consonant with the lack of selection in the whole science-fiction field). Unfortunately, the work of Dick praised above also has its reverse side. One is used to calling such work uneven. The contradictions in *The Three Stigmata of Palmer Eldritch* and *Ubik* (and also partly in *Solar Lottery*) are of a fleeting nature. These seeming contradictions constitute the claim of completeness—the semantic value of the work (as I tried to show very briefly). Therefore the local contradictions are meaningful messages that direct the reader's attention to the problems that underlie the works. The novel *Galactic Pot-Healer* is only negligible. Every author is free to produce works of different value; there is no law against a great epic master allowing himself a novel of pure entertainment.

Our Friends from Frolix 8 and *Do Androids Dream of Electric Sheep?* are not unimportant literature, but they cheat the reader. Especially in the latter do we see the sad picture of an author who squanders his talent by using brilliant ideas and inspirations to keep up a game of cops and robbers. This is far worse than putting together a valueless whole from valueless parts. The idea of the "Penfield apparatus," with which one can arbitrarily change one's own mental disposition, is a brilliant one, but it does not play a role in the novel. In order to unravel the logical mystery that makes up *Do Androids Dream of Electric Sheep?* a whole study would be necessary, but it would have to be written with the embarrassed feeling that it is wholly superfluous.

But I must say this without furnishing proof. The first premise of the plot is that a policeman may kill on the spot everyone who is discovered to be an android, because on earth only androids kill their masters. (This premise does not hold good in the face of what is written later in the book.) We get to know that some androids do not know their true nature, because they have been filled with the incorrect information that they are normal humans. The police system has been undermined by androids who, disguised as humans, kill policemen in order to bear false witness that the dead human has been unmasked as an android. At the same time, we discover that some policemen have the same type of android nature—i.e., with an artificially implanted consciousness that they are humans. But if somebody does not know himself whether he is an "android replica" or a normal policeman, in what sense is this "infiltration"? If an android has a synthetically "humanized" consciousness with a falsified memory, for what is he called to account? How can one be responsible for that which one has no knowledge of? With these actions did Dick intend to present a model of discrimination, such as the kind of persecution of the Jews administered under the label "final solution"? But then (1) the androids are innocent victims and should not be depicted as insidious creatures, something that the novel does in places, and (2) people who are persecuted—e.g., persecuted because of their race—are certainly conscious of their innocence but at the same time conscious of their identity, which is not the case with the androids. In other ways the parallel is not valid. It remains obscure whether every android is killed on the spot because of what he once did (he is supposed to have killed his master) or because of what he is. As I have shown, the claim that every android is a murderer because it is unthinkable there is an android without an owner is not valid. Why are there no humans, masters of androids, who die natural deaths in their beds?

As for the difference between human and android, we hear that it is almost impossible to distinguish between humans and androids with one hundred percent accuracy. To do this one needs a psychological test that measures the suspect's reactions with a psychogalvanic apparatus. The test is nonsense; besides, on another occasion we hear that androids have a life span of only a few years, since the cells of their tissue cannot multiply. Therefore it is not child's play to discover the difference by means of an organic examination of a microscope slide preparation of cell tissue, a procedure that takes about three minutes.

There is no unequivocal answer to all these questions. Situations to shock the readers must be multiplied at all costs. A trial to identify a suspect is far less shocking than the situation in which two policemen, working hand in glove, may kill one another if either of them should suddenly be unmasked as an android. This is all the more thrilling if neither of them, subjectively, knows who he really is, android or human. Then, both are subjectively innocent, both could be androids, or only one, or none— all of which heightens the tension, but at the same time increases the nonsense. In order to shock us when applied, the differentiating test must be applied fast and surely, but then suspense is lost if it is not coupled with the uncertainty of whether the suspect is an android or not, but with uncertainty of whether the test itself might fail, which causes somebody's death instantly, in error. Because the author did not want to do without these logically exclusive alternatives, the test must be at the same time reliable and unreliable, the androids must act at the same time with malice aforethought and in complete innocence; as an android one is at the same time conscious and unconscious of one's nature; a girl who has slept with a policeman is sentenced to death because it is forbidden for androids to sleep with humans; however, at the same time the girl does not know she is an android, etc., ad lib. The problem that is spelled out originally and begins to unfold, of human conflict with humanlike creations endowed with spirit by humans themselves, is torn to shreds, while the game of cops and robbers continues merrily. This nonsense, offered by the author of *Ubik*, can be construed as an offense to the reader, an offense which, however, evaporates without trace in the highly concentrated thoughtlessness of the science-fiction milieu.

We cannot deny this: the author of *Ubik* knew quite well what he was doing. But did criticism catch him red-handed and hold him responsible? I do not jest: for he who could write *Ubik* must understand the fraudulent character of his work. Criticism only took offense at his novel for being, in a way, insipid—i.e., not as full of suspense as the best of Dick. Such a brew of trite remarks is held out as criticism in science fiction.[7]

---

7. A lack of theoretical essays on science fiction was the reason for my career as a Robinson Crusoe; like the unhappy man on a desert island, I had to sweat for years, under the most primitive conditions, to produce the necessary (intellectual) tools by myself. My tactic concerning trash was to ridicule it—i.e., to blow up its model until its nonsense, multiplied many times, became ludicrous. But this is the simplest of tactics. On my own I thought there was no better way than to avoid trash and to remove all traces of it from my work.

Dick set me right, and for that reason—as a guidepost—his work is so important. With the tactics I was using I could write only humorous (or grotesque) works: this is worse than if one remains in earnest all the time. It is worse because humor shows up the rich ambiguity of an

There is no justification for this primitive dalliance; there is only an explanation, of a general character, which transcends the work itself. Ross Ashby proves that intelligence is a quality that does not foster survival under all possible variants of environments. In some environments stupidity serves better the drive for self-preservation. He spoke of rats; I would like to apply this claim to that part of literature called "science fiction." For in science fiction

---

earnest way of narration in but a lesser degree. The reader must recognize that an example has been ridiculed, or else the reader and writer are as much at cross-purposes as when somebody does not grasp the point of a joke; one cannot misunderstand a joke and savor it at the same time. Therefore humorous prose is assured of a more ready reception than complex prose that wants to be taken seriously. Because of Dick's method of "transformation of trash," I have found a third (just this) tactic of creation. A novel by Dick is not bound to be—and often is not—understood, because of its peculiar maximum span of meanings; because trash is not ridiculed; because the reader can enjoy its elements and see them isolated from reciprocal relationships within the same work. This is better for the work, since it can survive in different ways in the reader's environment, either correctly or incorrectly understood. Similarly, one can recognize a humorist at first glance, but not a man who makes use of Dick's tactics. It is far more difficult to grasp the complexity of the work in its entirety, and in no other way can we deal with the "transformation of trash."

Only a complete lack of a theory of science fiction makes it comprehensible why the New Wave of science fiction did not pick Dick as their guiding star. The New Wavers knew that they should look for something new, but they did not have the slightest idea what it could be. Surely there is no more diffuse definition of anything than that of the New Wave, which is supposed to be represented on the one hand by Spinrad, on the other by Delany, and on a third by Moorcock. Until now the New Wave has succeeded well in making science fiction quite boring, but this is the only characteristic in which it is approaching the state of modern prose in the Upper Realm. Repressed but powerful inferiority complexes are constantly at work, and we can detect this

what does it matter if *Ubik* is a piece of gold and *Do Androids Dream of Electric Sheep?* a counterfeit coin? I don't know what an average reader thinks while reading these two novels. If we could reproduce his thoughts as they correspond to his behavior as a library borrower, we must conclude that he has an extremely short memory; at the utmost he can remember what is printed on one page. Or he does not think at all; an alternative that scares me so much, however, that I'd prefer to drop it.

The problem remains that all science-fiction books are similar to one another—not according to their content, but according to the way they are received. Innumerable

---

because all the experimenters seem to believe from the bottoms of their hearts that the medicine and models for redeeming science fiction can be found only in the Upper Realm. Out of this belief came Farmer's *Riders of the Purple Wage* (no mean piece of prose, but of a markedly secondary, or even tertiary, character to Farmer's model, Joyce's *Ulysses,* which is itself modeled on *The Odyssey*) and *Stand on Zanzibar,* which, as we all know, was written by Brunner on the model of *Manhattan Transfer* by Dos Passos. The New Wavers seized expressionism, surrealism, etc., and so they completed a collection of old hats; it becomes a race backward which still arrives in the nineteenth century before they know it. But a blind search can give only blind results; just "blind shells" (duds).

As I said, I believe that a writer can either make a caricature of trash, and ridicule it, or throw it away. Dick found out how to blaze a third trail, a discovery that was important not just for himself, but that remained unnoticed. The newness of Ursula K. Le Guin's *The Left Hand of Darkness* was observed instantly because it is localized in the action, but the more volatile discovery by Dick was misjudged because it cannot be localized and can be described only with the utmost difficulty for the reasons I have set out. It is not sufficient, milords critics, to enjoy a book, and criticism is not a cry of joy; one must not only know how to prove that one was delighted but also know how to explain by what one was delighted and charmed.

imitations of each original work appear, so that the originals are buried beneath mountains of trash, like cathedral towers around which garbage has been dumped for so long that only the spires project out of the rubbish that reaches toward heaven. In this context the question arises as to how many gifted beginners have insufficient power to preserve their individuality as writers—unless by way of compromise, like Dick—in spite of the equalizing trends of science fiction.

Probably the pressure of trivial literature has crushed many highly talented writers with the result that today they deliver the products that keep highbrow readers away from science fiction. This process brings about a negative selection of authors and readers: for even those writers who can write good things produce banalities wholesale: the banality repels intelligent readers away from science fiction; as they form a small majority in fandom the "silent majority" dominates the market, and the evolution into higher spheres cannot occur. Therefore, in science fiction, a vicious circle of cause and effect coupled together keeps the existing state of science fiction intact and going. The most intelligent and most demanding readers, who form a small minority, still long for "better" science fiction and feel ill at ease when reading its current production, showing their uneasiness in their letters of comment and essays in fanzines. The "normal" reader—i.e., the silent majority and their representatives in fanzines—gains the impression somehow that the others are tense, scurrilous, and even malicious creatures just like—I wrote something like this once in a private letter—missionaries in a whorehouse —i.e., people who feel that they are doing their duty but at the same time are conscious that their efforts at conversion are powerless and that they seem out of place. The missionaries, ready to make the greatest sacrifices, can just as little change a whorehouse into a temple as "genial" readers can change science fiction into a fully qualified citizen of the Upper Realm of Literature.

I'll close this essay with one last remark: the disfigurement of Dick's work is the price that he had to pay for his "science-fiction citizenship." Dick owes his exuberant growth, as well as his own peculiar downfalls, to this circle of life, which, like a dull teacher, cannot distinguish its brightest pupils from the plodding grinds. This circle of life, like such a teacher, strives to treat all its subordinates in the same way, a way improper in schools, and disastrous in literature.

APPENDIX

Ubik *as Science Fiction*

In *Science Fiction Commentary 17*, George Turner wrote: "In *Ubik* we are given the living and the half-living; the half-living are actually dead but exist in another version of reality until their vestigial remainders of consciousness finally drain away. Their "reality" is subject to manipulation by a strong personality among the half-living, which piles complexity on complexity, until inconsistencies begin to stand out like protest posters. The plotting is neat, but cannot override the paradoxes. The metaphor fails because it cannot stand against the weight of reality as we know it."

Now I am ready to prove that there is a rational viewpoint from which *Ubik* can be seen as a novel based on scientifically sensible notions. Here is the line of proof.

In *Ubik* dying people are put into a state of "half-life" if medicine does not know how to heal them. The critically ill are placed in "cold packs" in which their bodies are intensively cooled down. At a very low temperature, their life functions decelerate so that death cannot occur. This is not fantasy. We know today that at temperatures close to 0° Kelvin for all practical purposes the growth of cancer cells stops, and even deadly poisons no longer destroy cells. Therefore an analogue of the process mentioned in *Ubik* can be realized today, except that it would be regarded as senseless to carry it out. Although cooling (better known as hibernation) will delay death and stop agony, one cannot speak of saving the patient: he is unconscious, he cannot be allowed to be warmed up to consciousness again, because then the death that has been delayed will occur. People speak of freezing a man and preserving him in this state of cryogenics until medicine discovers a method of healing this special case after years or centuries. We do not know yet whether reversible cold death, the idea of which lies at the base of this opinion, can be realized, because until the present day, experiments performed on mammals have shown no positive results; freezing and later defreezing wreaks irreversible damage on all tissues. *Ubik* presupposes that reversible cold death *cannot* be realized—something considered by specialists to be plausible or even highly probable. Thus hibernation can be regarded as useless, and freezing at low temperatures as unobtainable. But there is one escape route, viz., one could keep the *body* of the patient in a state of continuous hibernation and supply his brain with warm blood with a suit-able apparatus (artificial heart and lungs), so that the patient will regain consciousness.

The patient would find himself in the same position as a paralytic, or maybe we should call it a situation much worse than that. His sense organs do not function, for only his brain can be supplied with blood; however, even if someone were ready to face such a cruel risk as near-death, even then he could not be helped. For we know that the idea of keeping intact the paraphysiological functions of an isolated brain is utopian. When the normal flux of sense data to the brain ceases, and a state of sensory deprivation sets in, an ever-increasing decay of all, espe-cially the higher, brain functions sets in. An isolated brain cannot function normally; therefore we meet a barrier even in this escape route.

But all is not yet lost: if we succeed in creating a syn-thetic environment for the patient's brain, he will continue to live, although not in our normal reality—he will live in a substitute reality. This pseudoreality is the common good (or bad, as you like) of all people in cold storage. The key question to answer is whether we can create a substi-tute world for those lying in cold storage, and if so, how? Now we cannot put into effect such an achievement at the moment, but the chances of doing so are quite good. Often during surgical operations on the brain the cerebral cortex has been irritated electrically and, circumstances permit-ting (with which I do not wish to deal here), this irritation may produce a series of hallucinations that the patient lives through as reality. The subject hears the voice of a dead acquaintance, sees him, witnesses whole scenes from his past, and so on. Please bear in mind that these are primitive experiments to which very little time was devoted, because the main purpose of the operation was to heal the patient, and one is not allowed to attempt tests that carry with them the slightest shadow of danger. Perhaps we will gain more knowledge, which will allow us to perfect this method. There must be machines, which we can call simulators or environment-producers, to which people lying in cold storage could be connected. The simulator becomes a source of information used necessarily to create a fictitious environment in the patient's brain; it works according to a program attuned to the needs of each case and becomes a fountain of new facts and impressions previously unknown to the patient. (Even today we can bring about by irritation of the cerebral cortex not only sensory hallucinations, but also feelings, including, for example, erotic experiences.)

In principle, the technical problem in the real world is soluble, and so we come to the next, untechnological, question: how much knowledge can the patient have about his true situation? *Ubik* makes the assumption that sane people in cold storage, such as Runciter's wife, have been conscious of their situation for years, but also some people, such as Joe Chip, who was put on ice after an accident, and those placed there because of incurable disease, do not know about their situation. Somebody—and this happens to Joe Chip—meets with catastrophe, loses consciousness, regains it after a period of time and finds himself returned to his well-known environment without knowing that it is part of a pseudoreality to which he is condemned "for life" because this is the only way to save him.

Morally it is quite questionable whether the false belief of these people that they are still living normal lives should be maintained—but this problem is irrelevant because a much more important one displaces it: i.e., his next-of-kin prefers the situation in which the patient lives to his death; though at the same time nobody could call it an agreeable situation. People are not content to keep the patient alive, because, from the point of view of people in the normal world, he is leading only a half-life isolated from the real world. They want to reach him, to talk to him, listen to him, etc. This is technically possible, but only under the most extraordinary conditions. Pseudoreality makes up an integral whole for the patient; therefore if someone who exists outside intrudes, the patient experiences this intrusion as an anomaly in his environment. The "quest" cannot reach into pseudoreality in a fully plausible and harmless way. This is important if a patient such as Runciter's wife is conscious of the situation. But it is extremely important if he or she does not know it—as in the case of Joe Chip.

Two curious phenomena must still be explained: (1) the "mad" behavior of pseudoreality, and (2) the manipulation by one man in cold storage of the consciousness of his fellow sufferers. (In *Ubik* the problem is the curious relationship formed among Emily, Runciter's wife, Joe Chip, and the strange man named Jorg.)

The first phenomenon is a realistic presentation of a fictitious technology. We may, in advance, claim that whichever way the technology of reality-fission will be realized, it must be subject to certain malfunctions because no technology is invulnerable to malfunctions. The fact that at some time a breakdown in the production of pseudoreality will occur can be regarded as a realistic prediction, since none of today's predictions can tell us what kind of mishaps will happen. *Ubik*'s author was justified in describing the "breakdowns" and "defects" of pseudoreality at his own discretion. Different types of disasters may occur.

In pseudoreality certain anomalies of the flow of time and space might happen, and both have a dreamlike character, i.e., they resemble what we experience in dreams. This type of creation of "reality breakdowns" seems to be correct insofar as (according to what we said before) the main source of the information that makes up pseudoreality is the brain of the man lying in cold storage; in this way we can account for the fact that each relaxation of the direction of psychic processes by the simulator correlates with changed appearances in the mind of the patient. He will experience this as a change of environment, as if in a dream. (At this point I should like to remark that as a rule a dream is not recognized as such by the dreamer; for this reason Joe Chip also does not think of such an interpretation of the events around him.)

We may assume that the "overgrowth" of one consciousness by another occurs because a lot of people are lying in cold storage and, for economic reasons, not everyone is allotted a separate simulator. Rather, a handful of people is always connected with a multichannel machine. Even if one circuit is insulated from the others, it may happen that electrical impulses flash across, or cause the induction of another current; subjectively, this may be experienced as the "devouring" of one consciousness by another, neighboring, one.

The last question to be answered is: who is really lying in cold storage: Runciter or Joe Chip? Because of *all* the facts found in *Ubik,* one may conclude that both men lie in cold storage—that all the men on the moon were killed by the explosion and subjected to cold-storage treatment.

Quod erat demonstrandum—and in several places we have "filled" the gaps left in the novel. But it would not be correct to speak in earnest about such "gaps."

First, an author need not necessarily describe the technological details in a novel. As is well known, writers of contemporary novels do not describe the principles that underlie the functions of refrigerators, radios, and cars, and in these novels we would look in vain for the information that all the main characters are "vertebrates" and "mammals." The basic assumption of *Ubik* is a technology of split reality, and it is not particularly important what kind of technology caused this split, so it need not be described in detail. It can occur in many ways; the technological details have secondary importance. The most important detail is that in a world where split reality has already been realized, its inhabitants face new, previously unknown dilemmas and must solve problems having the greatest impact. The existence of such a technology changes the ontological perspective of life and, as *Ubik* shows convincingly, the problem is not just that of people put in cold storage because they are severely injured. In principle, anyone can be incarcerated in a pseudoworld for his whole life. Whether this is legal or illegal is a problem of jurisprudence, not philosophy. In a world with split reality, general knowledge shows that, as well as the normal level of reality, other levels may exist, levels that may exist for some other people—or for everybody. As always, this is a question of the price to be paid for so-called progress (in *Ubik,* progress in the battle against death).

At any rate, the point set out above is a perspective from

which the novel may be seen as a science-fiction work that depicts the human consequences of a biotechnological revolution. Perhaps it is not superfluous to remark in the second place that observers who watch the spectacle of a highway catastrophe do not usually indulge in reflections that call into question the facts of civilization and the history of technology; when people are looking at destroyed cars and maimed bodies they do not think about the price that has been exacted in human lives because Otto once invented the four-stroke engine and other inventors put this motor into the body of an old coach. So we may doubt whether the above technological exegesis is really necessary and whether we may think that Dick should of his own accord fill the gaps in technological detail that I have tried to fill.

Rather, I believe that Dick left no gaps in the novel, and in fact that the technological explanation is superfluous. It pursued only one object: I wanted to demonstrate that the novel is coherent as science fiction as well and that contradictions and loose ends in its structure are not in question. If technological details abounded in *Ubik* they would interfere with our reading; they do not add anything relevant to the text, and they can only rationalize it in a way that the author does not like. From the point of view of an artist, he is correct, for this novel is not "futurological" science fiction, though it may be read as such. However, Dick has taken a different point of view: he renounces all "empirical justifications" and "scientific" foundations. Primarily *Ubik* is a poetic achievement; we may draw this conclusion from the fact that the biotechnological premise, as outlined above, could also be the basis of a novel whose factual details were impeccable but, despite all this, a blind shell as a work of art. The contradictions in *Ubik* need not be defended at all costs by appealing to technological authority. The novel has neither gaps nor signs of the author's negligence. The "contradictions" form a mode of expression that serves to expose to full daylight the messages that are stressed by affection and a special philosophy of life. In a word, they are metaphors that should not be examined for empirical content, even if that seems possible. As I could show, even if they withstand logical and scientific tests, this is not their main value as an experience that can be exchanged with the currency of practical knowledge.

This experience is called catharsis.

POSTSCRIPT

The laws of science fiction form a dynamic structure at a balance of flow. Translated into the language of a futurologist, there are long-term, complex trends. There is no hope that they will be reversed. However, there are real possibilities that these trends will creep gradually into the Upper Realm of Literature, because of the ongoing explosion of information. The premise of selection that filters values implies a filter of sufficient capacity. But even today the capacity of this filter—the critics—as a value-selecting system is overtaxed by the quantity of books on the market. Generally, one is unaware of this situation. Consequently, the career of each literary work reminds us less of a directed trajectory than of something that takes on the motion of a Brownian particle—i.e., order becomes chaos. From the viewpoint of a critical filter, this chaos is not perceived easily, because a selection process is still taking place. But the fact that it takes place at all is no longer due to the filtration of the whole quantity of all the works that come onto the market, but to the random collision between prominent books and prominent critics. Since the number of books flowing onto the market increases continually, in the course of time the books form a kind of umbrella—i.e., they form a shield against the critics—and they frustrate an encompassing selection, something the critics do not realize for a long time because they are still fishing the "best" titles out of the stream of the market. However, they do not see those books that, although they are just as good as the ones picked out, or even better, remain unknown to them.

Selection no longer encompasses the whole quantity of published material, and this cultural area converts itself into a blind lottery. But this lottery takes only a marginal part in the selection of values. In due course, we can see that true values in abundance can have the same effect as a devastating flood. If they abound, these values begin to destroy themselves because they block all the filters intended to select them. Thus the fate of literature as a whole can become quite the same as that of trivial literature. Perhaps culture itself will be drowned in the Great Flood of information.

Translated from the German
by Werner Koopmann

PHILIP K. DICK: A VISIONARY
AMONG THE CHARLATANS

No one in his right mind seeks the psychological truth about crime in detective stories. Whoever seeks such truth will turn rather to *Crime and Punishment.* In relation to Agatha Christie, Dostoevsky constitutes a higher court of appeal, yet no one in his right mind will condemn the English author's stories on this account. They have a right to be treated as the entertaining thrillers they are, and the tasks Dostoevsky set himself are foreign to them.

If anyone is dissatisfied with science fiction in its role as an examiner of the future and of civilization, there is no way to make an analogous move from literary oversim-plifications to full-fledged art, because there is no court of appeal for this genre. There would be no harm in this except that American science fiction, exploiting its exceptional status, lays claim to occupy the pinnacles of art and thought. One is annoyed by the pretentiousness of a genre that fends off accusations of primitivism by pleading its entertainment character and then, once such accusations have been silenced, renews its overweening claims. By being one thing and purporting to be another, science fiction promotes a mystification that, moreover, goes on with the tacit consent of readers and public. The development of interest in science fiction at American universities has, contrary to what might have been expected, altered nothing in this state of affairs. In all candor it must be said, though one risk perpetrating a crime *laesae Almae Matris,* that the critical methods of theoreticians of literature are inadequate in the face of the deceptive tactics of science fiction. It is not hard to grasp the reason for this paradox: if the only fictional works treating of problems of crime were like those of Agatha Christie, then to just what kind of books could even the most scholarly critic appeal in order to demonstrate the intellectual poverty and artistic mediocrity of the detective thriller? Qualitative norms and upper limits are established in literature by concrete works and not by critics' postulates. No mountain of theoretical lucubrations can compensate for the absence of an outstanding fictional work as a lofty model. The criticism of experts in historiography did not undermine the status of Sienkiewicz's *Trilogy,* since there was no Polish Leo Tolstoy to devote a *War and Peace* to the period of the Cossack and Swedish wars. In short, *inter caecos luscus rex*—where there is nothing first-rate, its role will be taken over by mediocrity, which sets itself facile goals and achieves them by facile means.

What the absence of such model works leads to is shown, more plainly than by any abstract discussions, by the change of heart that Damon Knight, both author and respected critic, expressed in *Science-Fiction Studies* #3. Knight declared himself to have been mistaken earlier in attacking books by van Vogt for their incoherence and irrationalism, on the ground that, if van Vogt enjoys an enormous readership, he must by that very fact be on the right track as an author, and that it is wrong for criticism to discredit such writing in the name of arbitrary values if the reading public does not want to recognize such values. The job of criticism is, rather, to discover those traits to which the work owes its popularity. Such words, from a man who struggled for years to stamp out tawdriness in science fiction, are more than the admission of a personal defeat—they are the diagnosis of a general condition. If even the perennial defender of artistic values has laid down his arms, what can lesser spirits hope to accomplish in this situation?

Indeed, the possibility cannot be ruled out that Joseph Conrad's elevated description of literature as rendering "the highest kind of truth to the visible universe" may become an anachronism—that the independence of literature from fashion and demand may vanish outside science fiction as well, and then whatever reaps immediate applause as a best seller will be identified with what is most worthwhile. That would be a gloomy prospect. The culture of any period is a mixture of that which docilely caters to passing whims and fancies and that which transcends these things—and may also pass judgment on them. Whatever defers to current tastes becomes an entertainment, which achieves success immediately or not at all, for there is no such thing as a magic show or a football game which, unrecognized today, will become famous a hundred years from now. Literature is another matter: it is created by a process of natural selection of values, which takes place in society and which does not necessarily relegate works to obscurity if they are *also* entertainment, but which consigns them to oblivion if they are *only* entertainment. Why is this so? Much could be said about this. If the concept of the human being as an individual who desires of society and of the world something more than immediate satisfactions were abolished, then the difference between literature and entertainment would likewise disappear. But since we do not as yet identify the dexterity of a conjurer with the personal expression of a relationship to the world, we cannot measure literary values by numbers of books sold.

But how does it ever happen that something which is less popular can, in the historical long run, hold its own against that which scores prompt successes and even contrives to silence its opponents? This results from the aforementioned natural selection in culture, strikingly similar to such selection in biological evolution. The changes by virtue of which some species yield place to others on the evolutionary scene are seldom consequences of great cataclysms. Let the progeny of one species outsurvive that of another by a margin of only one in a million, and by and by only the former species will remain alive—though the difference between the chances of the two is imperceptible at short range. So it is also in culture: books that in the eyes of their contemporaries are so alike as to be peers part company as the years go by; facile charm, being ephemeral, gives way at last to that which is more difficult to perceive. Thus regularities in the rise and decline of literary works come into being and give direction to the development of the spiritual culture of an age.

Nevertheless, there can be circumstances that frustrate this process of natural selection. In biological evolution the result will be retrogression, degeneration, or at the very least developmental stagnation, typical of populations isolated from the outside world and vitiated by inbreeding, since these are most lacking in the fruitful diversity that is guaranteed only by openness to all the world's influences. In culture an analogous situation leads to the emergence of enclaves shut up in ghettos, where intellectual production likewise stagnates because of inbreeding in the form of incessant repetition of the selfsame creative patterns and techniques. The internal dynamics of the ghetto may appear to be intense, but with the passage of years it becomes evident that this is only a semblance of motion, since it leads nowhere, since it neither feeds into nor is fed by the open domain of culture, since it does not generate new patterns or trends, and since, finally, it nurses the falsest of notions about itself, for lack of any honest evaluation of its activities from outside. The books of the ghetto assimilate themselves to one another, becoming an anonymous mass, while such surroundings thrust whatever is better downward toward the worse, so that works of differing quality meet one another halfway, as it were, in the leveling process forced upon them. In such a situation publishing success not only may, but must, become the sole standard of evaluation, since a vacuum of standards is impossible. Hence, where there are no ratings on the merits, these are replaced by ratings on a commercial basis.

Just such a situation reigns in American science fiction, which is a domain of herd creativity. Its herd character manifests itself in the fact that books by different authors become as it were different sessions of playing at one and the same game or various figures of the selfsame dance. It should be emphasized that, in literary culture as in natural evolution, effects become causes by virtue of feedback loops: the artistic-intellectual passivity and mediocrity of works touted as brilliant repel the more exigent authors and readers, so that the loss of individuality in science fiction is at once a cause and an effect of ghetto seclusion. In science fiction there is little room left for creative work that would aspire to deal with problems of our time without mystification, oversimplification, or facile entertainment: e.g., for work that would reflect on the place that reason can occupy in the universe, on the outer limits of concepts formed on earth as instruments of cognition, or on such consequences of contacts with extraterrestrial life as find no place in the desperately primitive repertoire of science-fiction devices (bounded by the alternative "we win"/"they win"). These devices bear much the same relation to serious treatment of problems of the kind mentioned as does the detective story to the problems of evil inherent in mankind. Whoever brings up the heavy artillery of comparative ethnology, cultural anthropology, and sociology against such devices is told that he is using cannon to shoot sparrows, since it is merely a matter of entertainment; once he falls silent, the voices of the apologists for the culture-shaping, anticipative, predictive, and mythopoeic role of science fiction are raised anew. Science fiction behaves rather like a conjurer pulling rabbits from a hat, who, threatened with a search of his belongings, pretends to think we are crazy to suggest this and indulgently explains that he is just performing tricks—after which we promptly hear that he is passing himself off in public for an authentic thaumaturge.

Is creative work without mystification possible in such an environment? An answer to this question is given by the stories of Philip K. Dick. While these stand out from the background against which they have originated, it is not easy to capture the ways in which they do, since Dick employs the same materials and theatrical props as other American writers. From the warehouse that has long since become their common property, he takes the whole threadbare lot of telepaths, cosmic wars, parallel worlds, and time travel. In his stories terrible catastrophes happen, but this, too, is no exception to the rule; lengthening the list of sophisticated ways in which the world can end is among the standard preoccupations of science fiction. But where other science-fiction writers explicitly name and delimit the source of the disaster, whether social (terrestrial or cosmic war) or natural (elemental forces of nature), the world of Dick's stories suffers dire changes for reasons that remain unascertainable to the end. People perish not because a nova or a war has erupted, not because of flood, famine, plague, draught, or sterility, not because the Martians have landed on our doorstep; rather, there is some inscrutable factor at work that is visible in its manifestations but not at its source, and the world behaves as if it has fallen prey to a malignant cancer, which through metastases attacks one area of life after another. This is, be it said forthwith, apposite as a castigation of historiographic diagnostics, since in fact humanity does not as a rule succeed in exhaustively or conclusively diagnosing the causes of the afflictions that befall it. It is sufficient to recall how many diverse and in part mutually exclusive factors are nowadays adduced by experts as sources of the crisis of civilization. And this, be it added, is also appropriate as an artistic presupposition, since literature that furnishes the reader with godlike omniscience about all narrated events is today an anachronism that neither the theory of art nor the theory of knowledge will undertake to defend.

The forces that bring about world debacle in Dick's books are fantastic, but they are not merely invented ad hoc to shock the readers. We will show this with the example of *Ubik,* a work which, by the way, can also be regarded as a fantastic grotesque, a "macabresque" with obscure allegorical subtexts, decked out in the guise of ordinary science fiction.

If, however, it is viewed as a work of science fiction proper, the contents of *Ubik* can be most simply summarized as follows:

Telepathic phenomena, having been mastered in the context of capitalistic society, have undergone commercialization like every other technological innovation. So businessmen hire telepathists to steal trade secrets from their competitors, and the latter, for their part, defend

themselves against this "extrasensory industrial espio-
nage" with the aid of "inertials," people whose psyches
nullify the "psi field" that makes it possible to receive
others' thoughts. By way of specialization, firms have
sprung up that rent out telepathists and inertials by the
hour, and the "strong man" Glen Runciter is the proprie-
tor of such a firm. The medical profession has learned how
to arrest the agony of victims of mortal ailments, but still
has no means of curing them. Such people are therefore
kept in a state of "half-life" in special institutions,
"moratoriums" ("places of postponement"—of death, obviously). If they merely rested there unconscious in their
icy caskets, that would be small comfort for their surviving kin. So a technique has been developed for maintaining
the mental life of such people in "cold-pac." The world
which they experience is not part of reality, but a fiction
created by appropriate methods. Nonetheless, normal
people can make contact with the frozen ones, for the
cold-sleep apparatus has means to this end built into it,
something on the order of a telephone.

This idea is not altogether absurd in terms of scientific
facts: the concept of freezing the incurably ill to await the
time when remedies for their diseases will be found has
already come in for serious discussion. It would also be
possible in principle to maintain vital processes in a person's brain when the body dies (to be sure, that brain
would rapidly suffer psychological disintegration as a consequence of sensory deprivation). We know that stimula-
tion of the brain by electrodes produces in the subject of
such an operation experiences indistinguishable from ordinary perceptions. In Dick we find a perfected extension

of such techniques, though he does not discuss this explicitly in the story. Numerous dilemmas arise here: should the "half-lifer" be informed of his condition? is it right to keep him under the illusion that he is leading a normal life?

According to *Ubik,* people who, like Runciter's wife, have spent years in cold sleep are well aware of the fact. It is another matter with those who, like Joe Chip, have come close to meeting with a violent end and have regained consciousness imagining that they have escaped death, whereas in fact they are resting in a moratorium. In the book, it must be admitted, this is an unclear point, which is however masked by another dilemma: if the world of the frozen person's experiences is a purely subjective one, then any intervention in that world from outside must be for him a phenomenon that upsets the normal course of things. So if someone communicates with the frozen one, as Runciter does with Chip, this contact is accompanied in Chip's experiences by uncanny and startling phenomena—it is as if waking reality were breaking into the midst of a dream "only from one side," without thereby causing extinction of the dream and wakening of the sleeper (who, after all, cannot wake up like a normal man because he is not a normal man). But, to go a step further, is not contact also possible between two frozen individuals? Might not one of these people dream that he is alive and well and that from his accustomed world he is communicating with the other one—that only the other person succumbed to the unfortunate mishap? This, too, is possible. And, finally, is it possible to imagine a wholly infallible technology? There can be no such thing. Hence certain perturbations may affect the subjective world of the frozen sleeper, to whom it will then seem that his environment is going mad—perhaps that in it even *time* is falling to pieces! Interpreting the events presented in this fashion, we come to the conclusion that all the principal characters of the story were killed by the bomb on the moon, and consequently all of them had to be placed in the moratorium, and from this point on the book recounts only their visions and illusions. In a realistic novel (but this is a *contradictio in adiecto*) this version would correspond to a narrative that, after coming to the demise of the hero, would go on to describe his life *after death*. The realistic novel cannot describe this life, since the principle of realism rules out such descriptions. If, however, we assume a technology that makes possible the half-life of the dead, nothing prevents the author from remaining faithful to his characters and following them with his narrative—into the depths of their icy dream, which is henceforward the only form of life open to them.

Thus it is possible to rationalize the story in the above manner—on which, however, I would not insist too seriously, and that for two reasons at once. The *first reason* is that to make the plot fully consistent along the lines sketched above is impossible. If all Runciter's people perished on the moon, then *who* transported them to the moratorium? Another thing that does not yield to any rationalization is the talent of the girl who by mental effort alone was able to alter the present by transposing causal nodes in a past already over and done with. (This takes place before the occurrence on the moon, when there are no grounds for regarding the represented world as the purely subjective one of any half-life character.) Similar misgivings are inspired by Ubik itself, "the Absolute in a spray can," to which we will devote attention a little later on. If we approach the fictional world pedantically, no case can be made for it, since it is full of contradictions. But if we shelve such objections and inquire instead after the overall meaning of the work, we will discover that it is close to the meanings of other books by Dick, for all that they seem to differ from one another. Essentially it is always one and the same world that figures in them—a world of elementally unleashed entropy, of decay that not only, as in our reality, attacks the harmonious arrangement of matter, but also even consumes the order of elapsing time. Dick has thus amplified, rendered monumental and at the same time monstrous, certain fundamental properties of the actual world, giving them dramatic acceleration and impetus. All the technological innovations, the magnificent inventions, and the newly mastered human capabilities (such as telepathy, which our author has provided with an uncommonly rich articulation into "specialties") ultimately come to nothing in the struggle against the inexorably rising floodwaters of Chaos. Dick's province is thus a "world of preestablished disharmony," which is hidden at first and does not manifest itself in the opening scenes of the novel; these are presented unhurriedly and with calm matter-of-factness, just so that the intrusion of the destructive factor should be all the more effective. Dick is a prolific author, but I speak only of those of his novels that constitute the "main sequence" of his works; each of these books (I would count among them: *The Three Stigmata of Palmer Eldritch, Ubik, Now Wait for Last Year,* and perhaps also *Galactic Pot-Healer*) is a somewhat different embodiment of the same dramatic principle—the conversion of the order of the universe to rack and ruin before our eyes. In a world smitten with insanity, in which even the chronology of events is subject to convulsions, it is only the people who preserve their normality. So Dick subjects them to the pressure of a terrible testing, and in his fantastic experiment only the psychology of the characters remains nonfantastic. They struggle bitterly and stoically to the end, like Joe Chip in the current instance, against the chaos pressing on them from all sides, the sources of which remain, actually, un-fathomable, so that in this regard the reader is thrown back on his own conjectures.

The peculiarities of Dick's worlds arise especially from the fact that in them it is waking reality that undergoes profound dissociation and duplication. Sometimes the dis-sociating agency consists of chemical substances (of the hallucinogenic type—thus in *The Three Stigmata of Palmer Eldritch*); sometimes in "cold-sleep technique" (as precisely in *Ubik*); sometimes (as in *Now Wait for Last Year*) in a combination of narcotics and "parallel worlds." The end effect is always the same: distinguishing between waking reality and visions proves to be impossible. The technical aspect of this phenomenon is fairly inessential—it does not matter whether the splitting of reality is brought about by a new technology of chemical manipula-tion of the mind or, as in *Ubik,* by one of surgical opera-tions. The essential point is that a world equipped with the means of splitting perceived reality into indistinguishable likenesses of itself creates practical dilemmas that are known only to the theoretical speculations of philosophy.

This is a world in which, so to speak, this philosophy goes out into the street and becomes for every ordinary mortal no less of a burning question than is for us the threatened destruction of the biosphere.

There is no question of using a meticulous factual book-keeping to strike a rational balance for the novel, by virtue of which it would satisfy the demands of common sense. We are not only forced to but we ought to at a certain point stop defending its "science-fictional nature" and for a *second reason,* so far unmentioned. The first reason was dictated to us simply by necessity: given that the elements of the work lack a focal point, it *cannot* be rendered consistent. The second reason is more essential: the impossibility of imposing consistency on the text compels us to seek its global meanings not in the realm of events themselves, but in that of their constructive principle, the very thing that is responsible for lack of focus. If no such meaningful principle were discoverable, Dick's novels would have to be called mystifications, since any work must justify itself either on the level of what it presents literally or on the level of deeper semantic content, not so much overtly present in, as summoned up by, the text. Indeed, Dick's works teem with non sequiturs, and any sufficiently sensitive reader can without difficulty make up lists of incidents that flout logic and experience alike. But —to repeat what was already said in other ways—what is inconsistency in literature? It is a symptom either of incompetence or else of repudiation of some values (such as credibility of incidents or their logical coherence) for the sake of other values.

Here we come to a ticklish point in our discussion, since the values alluded to cannot be objectively compared.

There is no universally valid answer to the question whether it is permissible to sacrifice order for the sake of vision in a creative work—everything depends on what kind of order and what kind of vision are involved. Dick's novels have been variously interpreted. There are critics—Sam Lundwall is one—who say that Dick is cultivating an "offshoot of mysticism" in science fiction. It is not, though, a question of mysticism in the religious sense, but, rather, of occult phenomena. *Ubik* furnishes some grounds for such a conclusion. Does not the person who ousts Ella Runciter's soul from her body behave like a "possessing spirit"? Does not he metamorphose into various incarnations when fighting with Joe Chip? So such an approach is admissible.

Another critic (George Turner) has denied all value in *Ubik,* declaring that the novel is a pack of conflicting absurdities—which can be demonstrated with pencil and paper. I think, however, that the critic should not be the prosecutor of a book, but its defender, though one not allowed to lie: he may only present the work in the most favorable light. And because a book full of meaningless contradictions is as worthless as one that holds forth about vampires and other monstrous revenants, and since neither of them touches on problems worthy of serious consideration, I prefer my account of *Ubik* to all the rest. The theme of catastrophe had been so much worked over in science fiction that it seemed to be played out until Dick's books became a proof that this had been a matter of frivolous mystification. For science-fictional endings of the world were brought about either by man himself—e.g., by unrestrained warfare—or by some cataclysm as extrinsic as it was accidental, which thus might equally well not have happened at all.

Dick, on the other hand, by introducing into the annihilation ploy—the tempo of which becomes more violent as the action progresses—instruments of civilization such as hallucinogens, effects such a commingling of the convulsions of technology with those of human experience that it is no longer apparent just what works the terrible wonders—a *deus ex machina* or a *machina ex deo,* historical accident or historical necessity. It is difficult to elucidate Dick's position in this regard, because in particular novels he has given mutually incongruent answers to this question. Appeal to transcendence appears now as a mere possibility for the reader's conjectures, now as a diagnostic near-certainty. In *Ubik,* as we have said, a conjectural solution which refuses to explain events in terms of some version of occultism or spiritualism finds support in the bizarre technology of half-life as the last chance offered by medicine to people on the point of death. But earlier, in *The Three Stigmata of Palmer Eldritch,* transcendental evil emanates from the titular hero—that is, by the way, rather low-grade metaphysics, being akin to hack treatments of "supernatural visitations" and "ghosts," and all that saves the thing from turning into a fiasco is the author's virtuosity as a storyteller. And in *Galactic Pot-Healer* we have to do with a fabulous parable about a sunken cathedral on some planet and about the struggle that takes place between Light and Darkness over raising it, so that the last semblance of literalness of events vanishes here. Dick is, so I instinctively judge, perfidious in that he does not give unambiguous answers to the questions provoked by reading him, in that he strikes no balances and explains nothing "scientifically," but instead just confounds things, not only in the plot itself but also with respect to a superordinated category: the literary convention within which the story unfolds. For all that *Galactic Pot-Healer* leans toward allegory, it does not adopt this position either unambiguously or definitively, and a like indeterminacy as to genre is also characteristic of other novels by Dick, perhaps to an even higher degree. We thus encounter here the same difficulty about genre placement of a work that we met with in the writing of Kafka.

It should be emphasized that the genre affiliation of a creative work is not an abstract problem of interest only to theorists of literature. It is an indispensable prerequisite to the reading of a work. The difference between the theorist and the ordinary reader reduces itself to the fact that the latter places the book he has read in a specific genre automatically, under the influence of his internalized experiences—in the same way that we employ our native language automatically, even when we do not know its morphology or syntax from specialized studies. The convention proper to a concrete genre becomes fixed with the passage of time and is familiar to every qualified reader; consequently, "everybody knows" that in a realistic novel the author cannot cause his hero to walk through closed doors, but can on the other hand reveal to the reader the content of a dream the hero has and forgets before he wakes up (although the one thing is as impossible as the other from a common-sense point of view). The convention of the detective story requires that the perpetrator of a crime be found out, while the convention of science fiction requires rational accounting for events that are quite improbable and even seemingly at odds with logic and experience. On the other hand, the evolution of literary genres is based precisely on violation of storytelling conventions which have already become static. So Dick's novels in some measure violate the convention of science fiction, which can be accounted to him as merit, because they thereby acquire broadened meanings having allegorical import. This import cannot be exactly determined; the indefiniteness that originates from this favors the emergence of an aura of enigmatic mystery about the work. What is involved is a modern authorial strategy, which some people may find intolerable, but which cannot be assailed with factual arguments, since the demand for absolute purity of genres is becoming nowadays an anachronism in literature. The critics and readers who hold Dick's "impurity" with respect to genre against him are fossilized traditionalists, and a counterpart to their attitude would be an insistence that prosaists should keep on writing in the manner of Zola and Balzac, and only thus. In the light of the foregoing observations one can understand better the peculiarity and uniqueness of the place occupied by Dick in science fiction. His novels throw many readers accustomed to standard science fiction into abiding confusion, and give rise to complaints, as naïve as they are wrathful, that Dick, instead of providing "precise explanations" by way of conclusion, instead of solving puzzles, sweeps things under the rug. In relation to Kafka, analogous objections would consist in demanding that *The Metamorphosis* should conclude with an explicit "entomological justification," making plain when and under what circumstances a normal man can turn into a bug, and that *The Trial* should explain just what Mr. K. is accused of.

Philip Dick does not lead his critics an easy life, since he does not so much play the part of a guide through his fantasmagoric worlds as he gives the impression of one lost in their labyrinth. He has stood all the more in need of critical assistance, but has not received it, and has gone on writing while labeled a "mystic" and thrown back entirely on his own resources. There is no telling whether or how his work would have changed if it had come under the scrutinies of genuine critics. Perhaps such change would not have been all that much to the good. A second characteristic trait of Dick's work, after its ambiguity as to genre, is its tawdriness, which is not without a certain charm, being reminiscent of the goods offered at county fairs by primitive craftsmen who are at once clever and naïve, possessed of more talent than self-knowledge. Dick has as a rule taken over a rubble of building materials from the run-of-the-mill American professionals of science fiction, frequently adding a true gleam of originality to already worn-out concepts and, what is surely more important, erecting with such material constructions truly his own. The world gone mad, with a spasmodic flow of time and a network of causes and effects that wriggles as if nauseated, the world of frenzied physics, is unquestionably his invention, being an inversion of our familiar standard according to which only we, but never our environment, may fall victim to psychosis. Ordinarily, the heroes of science fiction are overtaken by only two kinds of calamities: the social, such as the "infernos of police-state tyranny," and the physical, such as catastrophes caused by nature. Evil is thus inflicted on people either by other people (invaders from the stars are merely people in monstrous disguises) or by the blind forces of matter.

With Dick the very basis of such a clear-cut articulation of the proposed diagnosis comes to grief. We can convince ourselves of this by putting to *Ubik* questions of the order just noted. Who was responsible for the strange and terrible things that happened to Runciter's people? The bomb attack on the moon was the doing of a competitor, but of course it was not in his power to bring about the collapse of time. An explanation appealing to the medical cold-pac technology is, as we have pointed out, likewise incapable of rationalizing everything. The gaps that separate the fragments of the plot cannot be eliminated, and they lead one to suspect the existence of some higher-order necessity, which constitutes the destiny of Dick's world. Whether this destiny resides in the temporal sphere or beyond it is impossible to say. When one considers to what an extent our faith in the infallible beneficence of technical progress has already waned, the fusion Dick envisages between culture and nature, between the instrument and its basis, by virtue of which it acquires the aggressive character of a malignant neoplasm, no longer seems merely sheer fantasy. This is not to say that Dick is predicting any concrete future. The disintegrating worlds of his stories—inversions, as it were, of Genesis, order returning to chaos—are not so much the future foreseen, as future shock, not straightforwardly expressed but embodied in fictional reality; an objectivized projection of the fears and fascinations proper to the human individual in our times.

It has been customary to identify the downfall of civilization falsely and narrowly with regression to some past stage of history—even to the caveman or downright animal stage. Such an evasion is often employed in science fiction, since inadequacy of imagination takes refuge in oversimplified pessimism. Then we are shown the remotest future as a lingering state of feudal, tribal, or slaveholding society, inasmuch as atomic war or invasion from the stars is supposed to have hurled humanity backward, even into the depths of a prehistoric way of life. To say of such works that they advocate the concepts of some cyclic (e.g., Spenglerian) philosophy of history would amount to maintaining that a motif endlessly repeated by a phonograph record represents the concept of some sort of "cyclic music," whereas it is merely a matter of a mechanical defect resulting from a blunt needle and worn grooves. So works of this sort do not pay homage to cyclic historiosophy, but merely reveal an insufficiency of sociological imagination, for which the atomic war or the interstellar invasion is only a convenient pretext for spinning out interminable sagas of primordial tribal life under the pretense of portraying the farthest future. Nor is it possible to hold that such books promulgate the "atomic credo" of belief in the inevitability of a catastrophe that will soon shatter our civilization, since the cataclysm in question amounts to nothing but an excuse for shirking more important creative obligations.

Such expedients are foreign to Dick. For him, the development of civilization continues, but is, as it were, crushed by itself, becoming monstrous at the heights of its achievement—which, as a prognostic viewpoint, is more original than the assuredly unilluminating thesis that, if technical civilization breaks down, people will be forced to get along by returning to primitive tools, even to bludgeons and flints.

Alarm at the impetus of civilization finds expression nowadays in the slogans of a "return to nature" after smashing and discarding everything "artificial," i.e., science and technology. These pipe dreams turn up also in science fiction. Happily for us, they are absent in Dick. The action of his novels takes place in a time when there can no longer be any talk of returning to nature or of turning away from the "artificial," since the fusion of the natural with the artificial has long since become an accomplished fact.

At this point it may be worthwhile to point out the dilemma encountered by futuristically oriented science fiction. According to an opinion quite generally held by readers, science fiction ought to depict the world of the fictional future no less explicitly and intelligibly than a writer such as Balzac depicted the world of his own time in *The Human Comedy.* Whoever asserts this fails to take into account the fact that there exists no world beyond or above history and common to all eras or all cultural formations of mankind. That which, like the world of *The Human Comedy,* strikes us as completely clear and intelligible is not an altogether objective reality, but is only a particular interpretation (of nineteenth-century vintage and hence close to us) of a world classified, understood, and experienced in a concrete fashion. The familiarity of Balzac's world thus signifies nothing more than the simple fact that we have grown perfectly accustomed to this account of reality and that consequently the language of Balzac's characters, their culture, their habits and ways of satisfying spiritual and bodily needs, and also their attitude toward nature and transcendence seem to us transparent. However, the movement of historical changes may infuse new content into concepts thought of as fundamental and fixed, as for example the notion of "progress," which according to nineteenth-century attitudes was equivalent to a confident optimism, convinced of the existence of an inviolable boundary separating what is harmful to man from what benefits him. Currently we begin to suspect that the concept thus established is losing its relevance, because the harmful ricochets of progress are not incidental, easily eliminated, adventitious components of it, but are, rather, gains achieved at such cost as, at some point along the way, to liquidate all the gain. In short, absolutizing the drive toward "progress" could prove to be a drive toward ruin.

So the image of the future world cannot be limited to adding a certain number of technical innovations, and meaningful prediction does not lie in serving up the present larded with startling improvements or revelations in lieu of the future.

The difficulties encountered by the reader of a work placed in a remote historical period are not the result of any arbitrariness on the writer's part, any predilection for "estrangements," any wish to shock the reader or to lead him up the garden path, but are an ineradicable part of such an artistic undertaking. Situations and concepts can be understood only through relating them to ones already known, but when too great a time interval separates people living in different eras there is a loss of the basis for understanding in common life experiences, which we unreflectingly and automatically imagine to be invariant. It follows that an author who truly succeeded in delineating an image of the far future would not achieve literary success, since he would assuredly not be understood. Consequently, in Dick's stories a truth value can be ascribed only to their generalized basis, which can be summed up more or less as follows: when people become ants in the labyrinths of the technosphere they themselves have built, the idea of a return to nature not only becomes utopian but cannot even be meaningfully articulated, because no such thing as a nature that has not been artificially transformed has existed for ages. We today can still talk of a return to nature, because we are relics of it, only slightly modified in biological respect within civilization, but try imagining the slogan "return to nature" uttered by a robot. Why, it would mean turning into deposits of iron ore!

The impossibility of civilization's returning to nature, which is simply equivalent to the irreversibility of history, leads Dick to the pessimistic conclusion that looking far into the future becomes such a fulfillment of dreams of power over matter as converts the ideal of progress into a monstrous caricature. This conclusion does not inevitably follow from the author's assumptions, but it constitutes an eventuality that ought *also* to be taken into account. By the way, in putting things thus, we are no longer summarizing Dick's work, but are giving rein to reflections about it, for the author himself seems so caught up in his vision that he is unconcerned about either its literal plausibility or its nonliteral message. It is the more unfortunate that criticism has not brought out the intellectual consequences of Dick's work and has not indicated the prospects inherent in its possible continuation, prospects and consequences advantageous not only for the author but also for the entire genre, since Dick has presented us not so much with finished *accomplishments* as with fascinating *promises*. It has, indeed, been just the other way around—criticism inside the field has instinctively striven somehow to domesticate Dick's creations, to restrain their meanings, emphasizing what in them is *similar* to the rest of the genre, and saying nothing about what is different—insofar as it did not simply denounce them as worthless for that difference. In this behavior a pathological aberration of the natural selection of literary works is emphatically apparent, since this selection ought to separate workmanlike mediocrity from promising originality, not lump these together, for such a "democratic" proceeding in practice equates the dross to the good metal.

Let us admit, however, that the charms of Dick's books are not unalloyed, so that it is with them somewhat as it is with the beauty of certain actresses, whom one had better not inspect too carefully at close range, on pain of being sadly disillusioned. There is no point in estimating the futurological likelihood of such details in this novel as those apartment and refrigerator doors that the tenant is forced to argue with. These are fictional ingredients created for the purpose of doing two jobs at once: to introduce the reader into a world decidedly different from the present-day one, and to convey a certain message to him by means of this world.

Every literary work has two components in the above sense, since every one exhibits a given factual world and says something by means of that world. Yet in different genres and different works the ratio between the two components varies. A realistic work of fiction contains a great deal of the first component and very little of the second, as it portrays the real world, which in its own right—that is, outside the book—does not constitute any sort of message, but merely exists and flourishes. Nevertheless, because the author makes, of course, particular choices when writing a literary work, these choices give it the character of a statement addressed to the reader. In an allegorical work there is a minimum of the first component and a maximum of the second, seeing that its world is in effect an apparatus signaling the actual content—the message—to the receiver. The tendentiousness of allegorical fiction is usually obvious, that of the realistic kind more or less well concealed. There are no works whatsoever without tendentiousness; if anyone speaks of such, what he actually has in mind is works devoid of expressly emphasized tendentiousness, which cannot be "translated" into the concrete credo of a world view. The aim of the epic, for example, is precisely to construct a world that can be interpreted in a number of ways—as the reality outside of literature can be interpreted in a number of ways. If, however, the sharp tools of criticism (of the structural kind, for instance) are applied to the epic, it is possible to detect the tendentiousness hidden even in such works, because the author is a human being and by that token a litigant in the existential process; hence complete impartiality is unattainable for him.

Unfortunately, it is only from realistic prose that one can appeal directly to the real world. Therefore, the bane of science fiction is the desire—doomed from the start to failure—to depict worlds intended at one and the same time to be products of the imagination and to signify nothing—i.e., not to have the character of a message but to be, as it were, on a par with the things in our environment, from furniture to stars, as regards their objective self-sufficiency. This is a fatal error lodged at the roots of science fiction, because where deliberate tendentiousness is not allowed, involuntary tendentiousness seeps in. By tendency we mean a partisan bias, or point of view, which cannot be divinely objective. An epic may strike us as just that objective, because the *how* of its presentation (the viewpoint) is for us imperceptibly concealed under the *what;* the epic, too, is a partisan account of events, but we do not notice its tendentiousness because we share its bias and cannot get outside it. We discover the bias of the epic centuries later, when the passage of time has transformed the standards of "absolute objectivity" and we can perceive, in what passed for a truthful report, the manner in which "truthful reporting" was at one time understood. There are no such things as truth or objectivity in the singular; both of these contain an irreducible coefficient of historical relativity. Now, science fiction can never be on a par with the epic, since *what* the science-fiction work presents belongs to one time (most often the future), whereas *how* it tells its story belongs to another time, the present. Even if imagination succeeds in rendering plausible *how* it might be, it cannot break completely with the way of apprehending events that is peculiar to the here and now. This way is not only an artistic convention; it is considerably more—a type of classification, interpretation, and rationalization of the visible world that is peculiar to an era. Consequently the problem content of an epic can be deeply hidden, but that of science fiction must be legible, otherwise the story, declining to deal with nonfictional problems and not achieving epic objectivity, slides fatally down and comes to rest on some such support as the stereotype of the fairy tale, the adventure thriller, the myth, the framework of the detective story, or some hybrid as eclectic as it is trashy. A way out of the dilemma may consist in works for which componential analysis, designed to separate what is "factual" from what forms the "message" ("seen" from a "viewpoint"), proves altogether impracticable. The reader of such a work does not know whether what he is shown is supposed to exist like a stone or a chair, or whether it is supposed also to signify something beyond itself. The indeterminacy of such a creation is not diminished by its author's commentaries, since the author can be mistaken in these, like a man who tries to explain the real meaning of his own dreams. Hence I consider Dick's own comments to be inessential to the analysis of his works.

At this point we might embark on an excursus about the origin of Dick's science-fictional concepts, but let just one example from *Ubik* suffice: to wit, the name that figures as the title of the book. It comes from the Latin *ubique,* "everywhere." This is a blend (contamination) of two heterogeneous concepts: the concept of the Absolute as eternal and unchanging order which goes back to systematizing philosophy, and the concept of the "gadget"—the handy little device for use on appropriate everyday occasions, a product of the conveyer-belt technology of the consumer society, whose watchword is making things easy for people at whatever they do, from washing clothes to getting a permanent wave. This "canned Absolute," then, is the result of the collision and interpenetration of two styles of thought of different ages, and at the same time of the incarnation of abstraction in the guise of a concrete object. Such a procedure is an exception to the rule in science fiction and is Dick's own invention.

It is hardly possible to create, in the way just noted, objects that are empirically plausible or that have a likelihood of ever coming into existence. Accordingly, in the case of Ubik it is a matter of a poetic—i.e., metaphorical —device and not of any "futurological" one. Ubik plays an important part in the story, emphasized still more by the "advertisements" for it that figure as epigraphs to each chapter. Is it a symbol, and, if so, of just what? This is not easy to answer. An Absolute conjured out of sight by technology, supposed to save man from the ruinous consequences of Chaos or Entropy much as a deodorant shields our sense of smell from the stench of industrial effluents, is not only a demonstration of a tactic typical nowadays (combating, for example, the side effects of one technology by means of another technology); it is an expression of nostalgia for a lost ideal kingdom of untroubled order, but also an expression of irony, since this "invention" of course cannot be taken seriously. Ubik moreover plays in the novel the part of its "internal micromodel," since it contains *in nuce* the whole range of problems specific to the book, those of the struggle of man against Chaos, at the end of which, after temporary successes, defeat inexorably awaits him. The Absolute canned as an aerosol, which saves Joe Chip at the point of death—though only for the time being: will this, then, be a parable and the handwriting on the wall for a civilization that has degraded the Sacred by stuffing it into the Profane? Pursuing such a train of associations, *Ubik* could finally be seen as a take-off on Greek tragedy, with the role of the ancient heroes, who strive vainly against Moira, assigned to the staff telepathists under the command of a big-business executive. If *Ubik* was not actually undertaken with this in mind, it in any case points in such a direction.

The writings of Philip Dick have deserved at least a better fate than that to which they were destined by their birthplace. If they are neither of uniform quality nor fully realized, still it is only by brute force that they can be jammed into that pulp of materials, destitute of intellectual value and original structure, that makes up science fiction. Its fans are attracted by the worst in Dick—the typical dash of American science fiction, reaching to the stars, and the headlong pace of action moving from one surprise to the next—but they hold it against him that, instead of unraveling puzzles, he leaves the reader at the end on the battlefield, enveloped in the aura of a mystery as grotesque as it is strange. Yet his bizarre blendings of hallucinogenic and palingenetic techniques have not won him many admirers outside the ghetto walls, since there readers are repelled by the shoddiness of the props he has adopted from the inventory of science fiction. Indeed, these writings sometimes fumble their attempts; but I remain after all under their spell, as often happens at the sight of a lone imagination's efforts to cope with a shattering superabundance of opportunities—efforts in which even a partial defeat can resemble a victory.

Translated from the Polish
by Robert Abernathy

THE TIME-TRAVEL STORY AND RELATED
MATTERS OF SCIENCE-FICTION
STRUCTURING

Let's look at a couple of simple sentences that logic, by virtue of a "disconnected middle" or by virtue of a tautology, asserts are always true, and let's investigate whether there can be worlds in which their veracity ceases. The first will be the ever real disjuncture: "John is the father of Peter or John is not the father of Peter." Any logician would acknowledge that this disjuncture satisfies at all times the requirement for truth, since *tertium non datur,* it is impossible to be forty percent father and sixty percent nonfather.

Next, let's work with a complex sentence: "If Peter has sexual relations with his mother, then Peter commits incest." The implication is a tautological one, since, according to the semantic rules of language, to have sexual relations with one's mother is tantamount to committing incest. (Our conjunction is not a complete tautology, since incest constitutes a concept broader than sexual relations with a mother, referring, rather, to relations with any person of such close kinship. We could bring the sentence to a perfect tautology, but this would necessitate complexities that would in no way alter the essence of the matter and merely make the argumentation more difficult.)

To simplify matters we shall investigate first the impact of changes on the veracity or falsity of the statement "John is the father of Peter." We should point out that what is involved here is a truly causative biological relation to the birth of a child, and not the ambiguous use of the designation "father" (since it is indeed possible to be a biological father and not be a baptismal father, or, conversely, to be a godfather, but not a parent).

Suppose John is a person who died three hundred years ago, but whose reproductive cells were preserved by refrigeration. A woman fertilized by them will become Peter's mother. Will John then be Peter's father? Undoubtedly.

But then suppose the following: John died and did not leave reproductive cells, but a woman asked a genetic technician to make up in the laboratory a spermatozoon of John from a single preserved cell of John's epithelium (all the cells of the body having the same genetic composition). Will John, once fertilization is complete, now also be Peter's father?

Now suppose the following case: John not only died,

but also did not leave a single bodily cell. Instead, John left a will in which he expressed the desire that a genetic technician perform the steps necessary to enable a woman to become the mother of a child of John—i.e., that such a woman give birth to a child and that the child be markedly similar to John. In addition, the genetic technician is not permitted to use any spermatozoa. Rather, he is supposed to cause a parthenogenetic development of the female ovum. Along with this he is supposed to control the genic substance and direct it by embryogenetic transformations in such a way that the Peter born is "the spit and image of John" (there are photographs of John available, a recording of his voice, etc.). The geneticist "sculptures" in the chromosomal substance of the woman all the features John craved for in a child. And thus, to the question "Is John the father or not the father of Peter?" it is now impossible to give an unequivocal answer of "yes" or "no." In some senses John is indeed the father, but in others he is not. An appeal to empiricism alone will not in itself furnish a clear answer. The definition will be essentially determined by the cultural standards of the society in which John, Peter's mother, Peter, as well as the genetic technician, all live.

Let's assume that these standards are fixed, and that the child realized in strict accordance with John's testamental instructions is generally acknowledged to be his child. If, however, the genetic technician, either on his own or at the instigation of others, made up forty-five percent of the genotypical features of the child not in accordance with the stipulations of the will, but in accordance with an entirely different prescription, it would then be impossible to maintain that John, in agreement with the standards of a given culture, either is or is not the child's father. The situation is the same as when some experts say about a picture reputed to be a work of Rembrandt: "This is a canvas by Rembrandt," whereas others say: "This is not a canvas by Rembrandt." Since it is quite possible that Rembrandt began the picture, but that some anonymous person finished the work, then forty-seven percent of the work could be said to originate from Rembrandt, and fifty-three percent from someone else. In such a situation of "partial authorship," *tertium datur.* In other words, there are situations in which it is possible to be a father only in part. (It is also possible to achieve such situations in other ways, e.g., by removing a certain number of genes from a spermatozoon of John and substituting another person's genes for them.)

The possibilities of the transformations mentioned above, which entail a change in the logical value of the disjunction—"John is the father of Peter or John is not the father of Peter"—lie, one may judge, in the bosom of a not too distant future. Thus a work describing such a matter would be fantastic today, but thirty or fifty years hence it might indeed be realistic. However, the work by no means needs to relate the story of a definite, concrete John, Peter, and mother of Peter. It could describe fictitious persons in a manner typical of any form of literary composition. The relational invariables between father, mother, and child would not have at that time the fictitious nature they have in the present. The invariables that concern paternity are today different from those of a time when genetic engineering would be realized. In this sense a composition written today and depicting a given situation without a "disconnected middle" in the predication of paternity may be considered a futurological prognosis or a hypothesis that may prove to be true.

For a real tautology to become a falsehood, the device of *travel in time* is necessary. Suppose Peter, having grown up, learns that his father was a very vile person—that he seduced Peter's mother and abandoned her only to disappear without a trace. Burning with the desire to bring his father to account for so despicable an act and unable to locate him in the present, Peter boards a time vehicle, sets out for the past and seeks out the father in the vicinity of the place where his mother was supposed to have resided at that time. The search, although very thorough, turns out to be in vain. However, in the course of establishing various contacts related to his expedition, Peter meets a young girl who attracts him. The two fall in love and a baby is conceived. Peter cannot remain permanently in the past, though; he is obliged to return to his old mother, for whom he is the sole support. Having been convinced by the girl that she has not become pregnant, Peter returns to the present. He has not succeeded in finding traces of his father. One day he finds in one of his mother's drawers a thirty-year-old photograph and to his horror recognizes in it the girl whom he loved. Not wishing to impede him, she committed a white lie, and hid her pregnancy. Peter thus comes to understand that he did not find his father for the simple reason that he himself is the father. So, Peter journeyed into the past to search for a missing father, assuming the name John to facilitate his search by remaining incognito. The upshot of this journey is his own birth. Thus, we have before us a circular causal structure. Peter is his own father, but, as against a superficial judgment, he did not commit incest at all, since, when he had sexual intercourse with her, his mother was not (and could not be) his mother. (From a purely genetic point of view, if we forget that—as is today believed—the causal circle is impossible, Peter is genotypically identical with his mother. In other words, Peter's mother for all practical purposes gave birth to him parthenogenetically, since, of course, no man inseminated her who was alien to her.)

This structure constitutes the so-called time loop, a causal structure characteristic of an enormous number of science-fiction compositions. The composition I described is a "minimal" loop, yet there is one still "smaller," created by Robert Heinlein in the story "All You Zombies" (1959).[1] Its plot is as follows: a certain young girl becomes pregnant by a man who then promptly disappears. She bears a child, or, more correctly, gives birth to it by Caesarean section. During the operation, the doctors ascertain that she is a hermaphrodite and it is essential (for reasons not explained by the author) to change her sex. She leaves the clinic as a young man who, because he was until quite recently a woman, has given birth to a child. She seeks her seducer for a long time, until it comes to light that *she herself* is he. We have the following circular situation: one and the same individual was in time T_1 both a girl and her partner, since the girl, transformed into a man by surgical intervention, was transferred by the nar-

---

1. The dates given in this essay are either for first publication, whether in serial or book form, or for serial/book publication. —R. D. Mullen.

rator to time T_1 from a future time, T_2. The narrator, a time traveler, "removed" the young man from time T_2 and transferred him to time T_1 so that the latter seduced "himself."

Nine months after time T_1 the child was born. The narrator stole this child and took it back in time twenty years, to moment T_0, so he could leave it under the trees of a foundling home. So the circle is completely closed: the same individual comprises "father," "mother," and "child." In other words, a person impregnated himself and gave birth to himself. The baby born as a result of this is left behind in time, bringing about in twenty years the growth of a girl who has in time T_1 sex with a young man from time T_2. The young man is she herself, transformed into a man by a surgical operation. The fact that a sexual hermaphrodite should not be able to bear a child is a relatively small hindrance, since the puzzling situation of a person's giving birth to himself is considerably "more impossible." What we are dealing with here is an act of *creatio ex nihilo.* All structures of the time-loop variety are internally contradictory in a causal sense. The contradictoriness is not, however, always as apparent as in Heinlein's story.

Frederic Brown writes about a man who travels into the past in order to punish his grandfather for tormenting his grandmother. In the course of an altercation he kills his grandfather before his father has been engendered. Thus the time traveler cannot then come into the world. Who, therefore, in fact killed the grandfather, if the murderer has not come into the world at all? Herein lies the contradiction. Sometimes an absent-minded scientist, having left something in the past, which he has visited, returns for the lost object and encounters his own self, since he has not returned exactly to the moment after his departure for the present, but to the time point at which he was before. When such returns are repeated, the individual is subject to multiple reproduction in the form of doubles. Since such possibilities appear to be pointless, in one of my stories about Ion Tichy (the "7th Journey"), I maximized "duplication" of the central character. Ion Tichy's spaceship finds itself in gravitational whirlpools that bend time into a circle, so that the spaceship is filled with a great number of different Ions.

The loop motif can be used, for instance, in the following ways. Someone proceeds into the past, deposits ducats in a Venetian bank at compound interest, and centuries later in New York demands from a consortium of banks payment of the entire capital, a gigantic sum. Why does he need so much money all of a sudden? So that he can hire the best physicists to construct for him a thus far nonexistent time vehicle, and by means of this vehicle go back in time to Venice, where he will deposit ducats at compound interest . . . (Mack Reynolds, "Compounded Interest" [1956]). Another example: in the future someone comes to an artist (in one story to a painter, in another to a writer) and gives him either a book dealing with painting in the future or a novel written in the future. The artist then begins to imitate this material as much as possible, and becomes famous, the paradox being that he is borrowing from his own self (since he himself was the author of that book or those pictures, "twenty years later").

We learn, further, from various works of this sort how

the Mesozoic reptiles became extinct thanks to hunters who organized a "safari into the past" (Frederic Brown), or how, in order to move in time in one direction, an equal mass must be displaced in the opposite direction, or how expeditions in time can reshape historical events. The latter theme has been used time and again, as in one American tale in which the Confederate States are victorious over the North (Ward Moore's *Bring the Jubilee* [1952, 1953]). The hero, a military historian, sets out for the past in order to investigate how the Southerners gained victory near Gettysburg. His arrival in a time machine throws General Lee's troop formations into disarray, which results in victory for the North. The hero is no longer able to return to the future, because his arrival also disturbed the causal chain upon which the subsequent construction of his time machine depended. Thus, the person who was supposed to have financed the construction of the machine will not do this, the machine will not exist, and the historian will be stuck in the year 1863 without the means to travel back into the original time. Of course here also there is an inherent paradox—just how did he reach the past? As a rule, the fun consists in the way the paradox is shifted from one segment of the action to another. The time loop as the backbone of a work's causal structure is thus different from the far looser motif of journeys in time per se; but, of course, it is merely a logical, although extreme, consequence of the general acceptance of the possibility of "chronomotion." There are actually two possible authorial attitudes, which are mutually exclusive: either one deliberately demonstrates causal paradoxes resulting from "chronomotion" with the greatest possible consistency, or else one cleverly avoids them. In the first instance, the careful development of logical consequences leads to situations as absurd as the one cited (an individual that is his very own father, that procreates himself), and usually has a comic effect (though this does not follow automatically).

Even though a circular causal structure may signalize a frivolous type of content, this does not mean that it is necessarily reduced to the construction of comic antinomies for the sake of pure entertainment. The causal circle may be employed not as the goal of the story, but as a means of visualizing certain theses, e.g., from the philosophy of history. Antoni Słonimski's story "Time Torpedo"[2] belongs here. It is a belletristic assertion of the *"ergo*ness" or *ergo*dicity of history: monkeying with events that have had sad consequences does not bring about any improvement of history; instead of one group of disasters and wars, there simply comes about another, in no way better.

A diametrically opposed hypothesis is incorporated into Ray Bradbury's "A Sound of Thunder" (1952). In an excellently written short episode, a participant in a "safari for tyrannosaurs" tramples a butterfly and a couple of flowers, and by that microscopic act causes such perturbances of causal chains involving millions of years, that upon his return the English language has a different orthography, and a different candidate—not liberal, but, rather, a kind of dictator—has won in the presidential election. It is only a pity that Bradbury feels obliged to set in motion complicated and unconvincing explanations to

---

2. Słonimski, born in 1895, was a Polish poet and essayist.

account for the fact that hunting for reptiles, which indeed fall from shots, disturbs nothing in the causal chains, whereas the trampling of a tiny flower does. (When a tyrannosaur drops to the ground, the quantity of ruined flowers must be greater than when the safari participant descends from a safety zone to the ground.) "A Sound of Thunder" exemplifies an "antiergodic" hypothesis of history, as opposed to Słonimski's story. In a way, however, the two are reconcilable: history can as a whole be "ergodic" if not very responsive to local disturbances, and at the same time such exceptional hypersensitive points in the causal chains can exist, the vehement disturbance of which produces more intensive results. In personal affairs such a "hyperallergic point" would be, for example, a situation in which a car attempts to pass a truck at the same time that a second car is approaching from the opposite direction.

As is usually the case in science fiction, a theme defined by a certain devised structure of occurrences (in this instance pertaining to a journey in time) undergoes a characteristic cognitive-artistic involution. We could have demonstrated this for any given theme, but let's take advantage of the opportunity at hand.

At first, authors and readers are satisfied by the joy of discerning the effects of innovations still virginal as far as their inherent contradictions are concerned. Then, an intense search is begun for initial situations that allow for the most effective exploitation of consequences that are potentially present in a given structure. Thus, the devices of chronomotion begin supporting, e.g., theses of history and philosophy (concerned with the "ergodicity" or nonergodicity of history). Then, grotesque and humorous stories like Frederic Brown's "The Yehudi Principle" (1944) appear: this short story is itself a causal circle (it ends with the words that it began with: it describes a test of a device for fulfilling wishes; one of the wishes expressed is that a story "write itself," which is what just happened).

Finally, the premise of time travel serves frequently as a simple pretext for weaving tales of sensational, criminal, or melodramatic intrigue; this usually involves the revival and slight refurbishment of petrified plots.

Time travel has been used so extensively in science fiction that it has been divided into separate subcategories. There is, for example, the category of *missent parcels* that find their way into the present from the future: someone receives a "Build-a-Man Set" box with "freeze-dried nerve preparations," bones, etc.; he builds his own double, and an "inspector from the future," who comes to reclaim the parcel, disassembles, instead of the artificial twin, the very hero of the story; this is William Tenn's "Child's Play" (1947). In Damon Knight's "Thing of Beauty" (1958) there is a different parcel—an automaton that draws pictures by itself. In general, strange things are produced in the future, science fiction teaches us (e.g., polka-dotted paint as well as thousands of objects with secret names and purposes not known).

Another category is *tiers in time.* In its simplest form it is presented in Anthony Boucher's "The Barrier" (1942), a slightly satiric work. The hero, traveling to the future, comes to a state of "eternal stasis," which, to protect its perfect stagnation from all disturbances, has constructed "time barriers" that foil any penetration. Now and then, however, a barrier becomes permeable. Rather disagreeable conditions prevail in this state, which is ruled by a police similar to the Gestapo (Stapper). One must be a slightly more advanced science-fiction reader to follow the story. The hero finds his way immediately into a circle of people who know him very well, but whom he does not know at all. This is explained by the fact that in order to elude the police he goes somewhat further back in time. He at that time gets to know these very people, then considerably younger. He is for them a stranger, but he, while he was in the future, has already succeeded in getting to know them. An old lady, who got into the time vehicle with the hero when they were fleeing from the police, meets as a result her own self as a young person and suffers a severe shock. It is clear, however, that Boucher does not know what to do with the "encountering oneself" motif in this context, and therefore makes the lady's shock long and drawn out. Further jumps in time, one after another, complicate the intrigue in a purely formal way. Attempts are begun to overthrow the dictatorial government, but everything goes to pieces, providing in the process sensationalism. *Antiproblematic escapism into adventure* is a very common phenomenon in science fiction; authors indicate its formal effectiveness, understood as the ingenious setting of a game in motion, as the skill of achieving uncommon movements, without mastering and utilizing the problematic and semantic aspects of such kinematics.

Such authors neither discuss nor solve the problems raised by their writing, but, instead, "take care" of them by dodges, employing patterns like the happy ending or the setting in motion of sheer pandemonium, a chaos that quickly engulfs loose meanings.

Such a state of affairs is a result of the distinctly "ludic" or playful position of writers; they go for an effect as a tank goes for an obstacle: without regard for anything incidental. It is as if their field of vision were greatly intensified and, simultaneously, also greatly confined. As in Tenn's story, the consequences of a "temporal lapse" in a postal matter are everything. Let us call such a vision *monoparametric*. At issue is a situation that is bizarre, amusing, uncanny, logically developed from a structural premise (e.g., from the presupposition of "journeys in time," which implies a qualitative difference in the world's causal structure). At the same time such a vision does not deal with anything more than that.

This can be seen readily from an example of "maximal intensification" of the subject of governments in time or "chronocracy," described by Isaac Asimov in his novel *The End of Eternity* (1955). "The Barrier" showed a single state isolating itself in the historical flow of events, as once the Chinese attempted to isolate themselves from disturbing influences by building the Great Wall (a spatially exact equivalent of a "time barrier"). *The End of Eternity* shows a government in power throughout humanity's entire temporal existence. Inspector-generals, traveling in time, examine the goings-on in individual epochs, centuries, and millennia, and by calculating the probability of occurrences and then counteracting the undesirable ones, keep in hand the entire system—"history extended in a four-dimensional continuum"—in a state of desirable equilibrium. Obviously, presuppositions of this sort are more thickly larded with antinomies than the plumpest pig is with bacon. While Asimov's great proficiency is manifested by the size of the slalom over which the narrative runs, it is, in the end, an ineffably naïve conception because no issues from philosophy or history are involved. The problem of "closed millennia," which the "tempocrats" do not have access to, is explained when a certain beautiful girl, whom an inspector falls in love with, turns out to be not a lowly inhabitant of one of the centuries under the dominion of the tempocracy, but a secret emissary from the "inaccessible millennia." The time dictatorship as a control over the continuum of history will be destroyed, and a liberated humanity will be able to take up astronautics and other select suitable occupations. The enigma of the inaccessible millennia is remarkably similar to the "enigma of the closed room" found in fairy tales and detective stories. The various epochs about which the emissaries of the chronocracy hover also recall separate rooms. *The End of Eternity* is an exhibition of formal entertainment to which sentiments about the fight for freedom and against dictatorship have been tacked on rather casually.

We have already spoken about the "minimal time loop." Let us talk now, simply for the sake of symmetry, about the "maximal" loops.

A. E. van Vogt has approached this concept in *The Weapon Shops of Isher* (1949, 1951), but let's expound it in our own way. As is known, there is a hypothesis (it can be found in Feynman's physics) that states that positrons are electrons moving "against the tide" in the flow of time. It is also known that in principle, even galaxies can arise from atomic collisions, as long as the colliding atoms are sufficiently rich in energy. In accordance with these presuppositions we can construct the following story: in a rather distant future a celebrated cosmologist reaches, on the basis of his own research as well as that of all his predecessors, the irrefutable conclusion that, on the one hand, the cosmos came into being from a single particle and, on the other, that such a single particle could not have existed. Where could it have sprung from? Thus he is confronted with a dilemma: the cosmos has come into being, but it could not come into being! He is horrified by this revelation, but, after profound reflection, suddenly sees the light: the cosmos exists exactly as mesons sometimes exist; mesons, admittedly, break the law of conservation, but do this so quickly that they do not break it. The cosmos exists on credit! It is like a debenture, a draft for material and energy which *must* be repaid immediately, because its existence is the purest one-hundred-percent liability both in terms of energy and in terms of material. Then, just what does the cosmologist do? With the help of physicist friends he builds a great "chronogun," which fires one single electron backward "against the tide" in the flow of time. That electron, transformed into a positron as a result of its motion "against the grain" of time, goes speeding through time, and in the course of this journey acquires more and more energy. Finally, at the point where it "leaps out" of the cosmos—i.e., in a place in which there had *as yet* been no cosmos—all the terrible energies it has acquired are released in that tremendously powerful explosion which brings about the universe! In this manner the debt is paid off. At the same time, thanks to the largest possible "causal circle," the existence of the cosmos is authenticated, and a person turns out to be the actual creator of that very universe! It is possible to complicate this story slightly; for example, by telling how certain colleagues of the cosmologist, unpleasant and envious people, meddled in his work, shooting on their own some lesser particles backward against the tide of time. These particles exploded inaccurately when the cosmologist's positron was producing the cosmos, and because of this that unpleasant rash came into being that bothers science so much today, namely, the enigmatic quasars and pulsars, which are not readily incorporated into the corpus of contemporary knowledge. These, then, are the "artifacts" produced by the cosmologist's malicious competitors. It would also be possible to tell how humanity both created and depraved itself, because some physicist shot the chronogun hurriedly and carelessly and a particle went astray, exploding as a nova in the vicinity of the solar system two million years ago, and damaging by its hard radiance the hereditary plasma of the original anthropoids, who therefore did not evolve into "man good and rational" as "should have happened" without the new particle. In other words, the new particle caused the degeneration of Homo sapiens—witness his history.

In this version, then, we created the cosmos only in a mediocre fashion, and our own selves quite poorly. Obviously a work of this sort, in whichever variant, becomes ironical, independently of its basic notion (i.e., the "self-creative" application of the "maximal time loop").

As one can see, what is involved here is an intellectual game, actually fantasy-making, which alters in a logical or pseudo-logical manner current scientific hypotheses. This is "pure" science fiction, or science fantasy, as it is sometimes called. It shows us nothing serious, but merely demonstrates the consequences of a reasoning that, operating within the guidelines of the scientific method, is used sometimes in unaltered form (in predicting the "composition percentage of paternity" we have in no way altered the scientific data), and sometimes secretly modified. And thus science fiction can be responsibly or irresponsibly plugged into the hypothesis-creating system of scientific thought.

The example of "self-creation" reveals first of all the "maximal proportions" of a self-perpetrating paradox: Peter gave birth only to himself, whereas in the universal variant, mankind concocted itself, and, what is more, perhaps not in the best manner, so that it would even be possible to use "Manichaean" terminology. Furthermore, this example at the same time demonstrates that the conceptual premise of essential innovations in the structure of the objective world presented is central to a science-fictional work (in the case of journeys in time, a change in causality is involved, by admitting the reversibility of that which we consider today as universally and commonly irreversible). The qualities of fictional material that serve a dominant concept are thus subject to an assessment based on their usefulness to this concept. Fictional material should in that case be an embodiment of a pseudo-scholarly or simply scholarly hypothesis—and that's all. Thus "pure" science fiction arises, appealing exclusively to "pure reason." It is possible to complicate a work with problems lying beyond the scope of such an intellectual game: when, for example, the "Manichaeism of existence" is interpreted as due to an error made by an envious physicist. Then an opportunity for sarcasm or irony arises as a harmonic "overtone" above the narrative's main axis. But by doing this, we have forced science fiction to perform "impure" services, because it is then not delivering scientific pseudo-revelations, but functioning in the same semantic substratum in which literature has normally operated. It is because of this that we call science fiction contaminated by semantic problems *"relational* science fiction."

However, just as "normal" literature can also perform high and low services—produce sentimental love stories and epics—relational science fiction shows an analogous amplitude. As was noted, it is possible to interpret it allegorically (e.g., Manichaeism in relation to the creation of the cosmos), and this will be the direction of grotesque or humorous departures from a state of "intellectual purity" that is somewhat analogous to "mathematical vacuity." It is also possible to overlay the history of creating the cosmos with melodrama, e.g., to make it part of a sensational, psychopathological intrigue (the cosmologist who created the universe has a wicked wife whom he nonetheless loves madly; or, the cosmologist becomes possessed; or also, faced with his deeds, the cosmologist goes insane and, as a megalomaniac, will be treated slightingly in an insane asylum, etc.).

Thus, in the end, the realistic writer is not responsible for the overall—e.g., the causal—structure of the real world. In evaluating his works, we are not centrally concerned with assessing the structure of the world to which they nonetheless have some relation.

On the contrary, the science-fiction writer is responsible both for the world in which he has placed his action and for the action as well, inasmuch as he, within certain limits, invents both one and the other.

Yet the invention of new worlds in science fiction is as rare as a pearl the size of a bread loaf. And so 99.9 percent of all science-fiction works follow, compositionally, a scheme, one of the thematic structures that constitute the whole science-fiction repertoire. For a world truly new in structural qualities is one in which the causal irreversibility of occurrences is denied, or one in which a person's individuality conflicts with an individual scientifically produced by means of an "intellectronic evolution," or one in which earthly culture is in communication with a non-earthly culture distinct from human culture not only nominally but qualitatively, and so forth. However, just as it is impossible to invent a steam engine, or an internal combustion engine, or any other already existing thing, it is also impossible to invent once more worlds with the sensational quality of "chronomotion" or of "a reasoning machine." Just as the detective story unweariedly churns out the same plot stereotypes, so does science fiction when it tells us of countless peripeties merely to show that by interposing a time loop they have been successfully invalidated (e.g., Thomas Wilson's "The Entrepreneur" [1952], which talks about the dreadful Communists having conquered the United States, and time travelers who start backward at the necessary point, invalidating such an invasion and dictatorship). In lieu of Communists, there may be Aliens or even the Same People Arriving from the Future (thanks to the time loop, anyone can battle with himself just as long as he pleases), etc.

If new concepts, those atomic kernels that initiate a whole flood of works, correspond to that gigantic device by which bioevolution was "invented"—i.e., to the constitutional principle of types of animals such as vertebrates and nonvertebrates, or fish, amphibians, mammals, and birds—then, in the "evolution of science fiction," the equivalent of type-creating revolutions were the ideas of time travel, of constructing a robot, of cosmic contact, of cosmic invasion, and of ultimate catastrophe for the human species. And, as within the organization of biological types a natural evolution imperceptibly produces distinctive changes according to genera, families, races, and so forth, so, similarly, science fiction persistently operates within a framework of modest, simply variational craftsmanship.

This very craftsmanship, however, betrays a systematic, unidirectional bias; as we stated and demonstrated, great concepts that alter the structure of the fictional world are a manifestation of a pure play of the intellect. The results are assessed according to the type of play. The play can also be "relational," involved with situations only loosely or not at all connected with the dominant principle. What connection is there, after all, between the existence of the cosmologist who created the world and the fact that he has a beautiful secretary whom he beds? Or, by what if not by a retardation device will the cosmologist be snatched away before he fires the "chronogun"? In this manner an idea lending itself to articulation in a couple of sentences (as we have done here) becomes a pretext for writing a long novel (where a "cosmos-creating" shot comes only in the epilogue, after some deliverers sent by the author have finally saved the cosmologist from his sorry plight). The purely intellectual concept is stretched thoroughly out of proportion to its inherent possibilities. But this is just how science fiction proceeds—usually.

On the other hand, rarely is a departure made from "emptiness" or "pure play" in the direction of dealing with a set of important and involved problems. In the world of science fiction it is structurally as possible to set up an adventure plot as a psychological drama; it is as possible to deal in sensational happenings as it is to stimulate thought by an ontological implication created by the narrative as a whole. It is precisely this slide toward easy, sensational intrigue that is a symptom of the degeneration of this branch of literature. An idea is permitted in science fiction if it is packaged so that one can barely see it through the glitter of the wrapping. As against conventions only superficially associated to innovations in the world's structure and which have worn completely threadbare from countless repetitions, science fiction should be stimulated and induced to deviate from this trend of development, namely, by involution away from the "sensational pole." Science fiction should not operate by increasing the number of blasters or Martians who impede the cosmologist in his efforts to fire the "chrono-gun"; such inflation is not appropriate. Rather, one should change direction radically and head for the opposite pole. After all, in principle the same bipolar opposition also prevails in ordinary literature, which also shuttles between cheap melodrama and stories with the highest aesthetic and cognitive aspirations.

It is difficult, however, to detect in science fiction any

improvement or outright redemption of this sort. An odd fate seems to loom heavily over its domain, which prompts writers with the highest ambitions and considerable talent, such as Ray Bradbury or J. G. Ballard, to employ the conceptual and rational tools of science fiction at times in an admittedly superb way, yet not in order to ennoble the genre, but, instead, to bring it toward an "optimal" pole of literature. Aiming in that direction, they are simultaneously, in each successive step, giving up the programmatic rationalism of science fiction in favor of the irrational; their intellect fails to match their know-how and their artistic talent. In practice, what this amounts to is that they do not use the "signaling equipment" of science fiction, its available accessories, to express any truly, intellectually new problems or content. They try to bring about the conversion of science fiction to the "creed of normal literature" through articulating, by fantastic means, such nonfantastic content, already old-fashioned, in an ethical, axiological, philosophical sense. The revolt against the machine and against civilization, the praise of the "aesthetic" nature of catastrophe, the dead-end course of human civilization—these are their foremost problems, the intellectual content of their works. Such science fiction is, as it were, a priori vitiated by pessimism, in the sense that anything that may happen will be for the worse.

Such writers proceed as if they thought that, should mankind acknowledge the existence of even a one-in-a-million or one-in-a-billion chance—transcending the already known cyclical pulsation of history, which has oscillated between a state of relative stabilization and of complete material devastation—such an approach would not be proper. Only in mankind's severe, resolute rejection of all chances of development, in complete negation, in a gesture of escapism or nihilism, do they find the proper mission of all science fiction that would not be cheap. Consequently they build on dead-end tragedy. This may be called into question not merely from the standpoint of optimism, of whatever hue and intensity. Rather, one should criticize their ideology by attempting to prove that they tear to shreds that which they themselves do not understand. With regard to the formidable movements that shake our world, they nourish the same fear of misunderstanding the mechanisms of change that every ordinary form of literature has. Isn't it clear what proportions their defection assumes because of this? Cognitive optimism is, first of all, a thoroughly nonludic premise in the creation of science fiction. The result is often extremely cheap, artistically as well as intellectually, but its principle is good. According to this principle, there is only one remedy for imperfect knowledge: better knowledge, because more varied knowledge. Science fiction, to be sure, normally supplies numerous surrogates for such knowledge. But, according to its premises, that knowledge exists and is accessible: the irrationalism of Bradbury's or Ballard's fantasy negates both these premises. One is not allowed to entertain any cognitive hopes—that becomes the unwritten axiom of their work. Instead of introducing into traditional qualities of writing new conceptual equipment as well as new notional configurations relying on intellectual imagination, these authors, while ridding themselves of the stigma of cheap and defective science fiction, in one fell swoop give up all that constitutes its cognitive value. Obviously, they are unaware of the consequences of such desertion, but this only clears them morally: so much the worse for literature and for culture, seriously damaged by their mistake.

Translated from the Polish
by Thomas H. Hoisington
and Darko Suvin

METAFANTASIA: THE POSSIBILITIES OF
SCIENCE FICTION

Let us demonstrate three possible types of science fiction by way of three fictional examples. Our first example is a work about a system for preventing earthquakes. It has recently been discovered that spraying water under high pressure into geological strata that lie under immense tectonic stress induces a series of harmless, microseismic movements in the earth. As the water penetrates the fissures of the deep-lying strata, it acts like lubricating oil, and helps the soil layers to slide away from one another. This preventive spraying facilitates the gradual reduction of tectonic stress, so that the pent-up seismic energies need not be released in the tremendous, destructive earthquakes that accompany the fierce shifts of geological formations. On the basis of this hypothesis, already verified in reality, one might write a science-fiction novel about the successful elimination of the catastrophes that threaten people living in earthquake-prone regions. This is a variation on the theme of "humanity's game with nature"; our stance concerning it would be unequivocal, since we would not have to re-evaluate or reorient our cultural norms to conclude that the prevention of earthquakes is a good thing and a worthwhile goal for which to strive.

Our second work might describe what happens when the use of a certain chemical that separates the sensations of pleasure from sex spreads throughout the earth. One possible rational motive for the use of the drug might be the desire to check the population explosion. Or there might also be a hostile motive: the drug might be a secret weapon in a covert military operation. It is not difficult to imagine the consequences. Since no one wants to indulge in sex of his own free will any more—after all, it is simply hard physical labor, totally devoid of pleasure—humanity is threatened with extinction. To prevent this disastrous eventuality, governments are forced to experiment with strategies for saving the human race. First, they try propaganda. But quickly they are forced to realize that the very same drawings, photographs, and movies whose distribution they had been obliged to prohibit not so long before no longer interest anyone at all now; on the contrary, they produce general disgust, since they are no more arousing for either sex than a picture of a washtub is for a washerwoman, or a photo of an ax for an exhausted woodcutter.

These seductive devices fail for a very simple reason:

once the act itself has lost its attraction, no amount of hinting, alluding, and suggesting can create a desire for it. Since the promotion of sex proves ineffective, governments resort to more pragmatic methods. They mobilize material incentives: rewards, premiums, decorations, extraordinary honors, social benefits, privileges, and magnificent titles with honorary diplomas. In the meantime, several industries go under: the cosmetics industry, part of the publishing industry (after all, who will read erotic literature when all it calls to mind is drudgery?), the film industry, as well as advertising—since they have been based on sex. The clothing and underwear industries are faced with a crisis greater than they have ever faced. Women's breasts now only remind people that humans are mammals; legs, that people can walk; and a painted mouth seems as bizarre as if someone had decided to glue an artificial ear on his bald head.

Naturally, researchers work feverishly to find an antidote that will neutralize the catastrophic effect of the drug; but in vain. As the new state of affairs stabilizes, new models of beauty emerge, models that provide security against every kind of erotic danger (for it can happen that one resigns himself to procreating to gain a medal or a title, only to find the potential partner repelled by the invitation; others may try to shirk their social responsibilities by making the illusion appear to be the reality, and thus supervisory committees are established to verify that everything is taking place as the social good demands; men declare that they deserve greater rewards because they have to put more work into it, while women protest that this is out of the question, and so forth). Under these conditions, perfect security lies in the companionship of someone visibly incapable of the sexual act (and so, not likely ever to suddenly demand it). Gray hair, potbellies, wheelchairs, and similar "antisexual" characteristics are accorded universal interest and respect as symbols of erotically disarmed paralysis.

A work of this sort would posit a certain anthropological hypothesis about the role of sexuality in the totality of human behavior.

The third example is of an entirely different order. It is a popular scientific book published in the mid-twenty-first century detailing the history of cosmological views, including the most recent theories. The author begins, naturally, at the beginning: long, long ago, humans, basing their thought on their relationships to their own products, conceived of the universe as an intentional object, like a pot or a table; there was a Someone who had created it, intentionally, and by design. The battle of ideas went on for centuries, until science appeared to establish that natural phenomena are not intentional objects. Thus, the trees, stones, atoms, clouds, oceans, rivers, and beyond them the planets, the sun, the stars, and the nebulae that constituted the objects of scientific inquiry were products of the natural processes of a heterogeneous evolution not conceived or designed by a personal being. Science discovered a series of objective regularities in these phenomena, and named them the fundamental laws of nature. Physics and astrophysics led the field, and the other branches of science queued up behind them.

But by the mid-twentieth century, theoretical views in the scientific world had come into grave conflict with one another. On the one hand, physics, planetology, astronomy, and evolutionary biology preached that the birth and development of life, which is crowned by the appearance of intelligent beings, is, in cosmic terms, normal, typical, average, and therefore a phenomenon belonging to the order of things. On the other hand, despite years of serious effort not a single trace had been discovered of any great, stellar-scale constructs that might have signaled the existence of a highly developed civilization, either in our own galaxy, or elsewhere. The persistence of this intolerable situation—produced by the contradiction between scientific expectations and the empirical data that had actually been gathered—swept the natural sciences, primarily biology and astronomy, into an ever-deepening crisis, until at last the inevitable ensued, and science resigned itself to the painful labor of restructuring its theoretical foundations.

Since we are here gathering in a nutshell something that itself amounts to a summary of an entire epoch's work (i.e., in our proposed popular scientific book) we cannot delve into the biographies of the learned people who set human thinking, including cosmogony and cosmology, on a completely new track. The first tentative hypotheses proposed by certain pioneering scientists were given the worst possible reception by the community of inquirers. But when the evidence of the "negative facts" became incontestable (i.e., the total absence of signs of "astrotechnical" constructs or traces), an extraordinary reversal occurred. Through their common efforts, scientists shaped new approaches and new models of the cosmos one after another, and the broad outlines of a new image of the universe began to unfold as follows.

Astrophysicists already know today that our sun and its planetary system belong to the so-called second stellar generation; the solar system is approximately five billion years old, while our whole galaxy is close to ten billion years old. Clearly then, the first generation of stars came into being before the formation of our solar system, in the remote mists of the cosmic past. With them came the planets, and on these planets life emerged. This was the first stage in the history of cosmic civilizations. When they attained a sufficiently high degree of scientific development they applied astrotechnics in an ever wider sphere of activity. For creatures at lower levels of development, the laws of nature are immutable attributes of being, but for those who have reached the higher planes of cognition, the laws of nature are no longer absolutely binding. Certain changes can be effected on them; the constant of gravitation, for example, can be reshaped, as well as the constants of electrical charges, the constant of maximum velocity, and so on. Since enormous distances separate the most developed civilizations from one another—distances of several hundred million light-years' magnitude, at the very least—they do not communicate with each other directly. They only infer the existence of their neighbors from certain observed facts: from certain gradual, noticeable changes in the laws of nature. Some of these transformations may benefit a given civilization, others may not. Therefore, each civilization approves and augments the former, and obstructs the latter, through its own astrotechnical activities. Thus begins the cosmogonic game played by the most developed civilizations of the universe.

This cosmogonic game is not military in nature, since

the partners do not use weapons and do not aim to annihilate one another. Rather, it is a co-operation justified by considerations beyond ethics: the annihilation or conquest of the partners would benefit no one, while by co-operating the partners help to sustain the trend of cosmogonic transformation most beneficial to everyone. Nor is the game a form of interstellar dialogue. Civilizations so advanced have nothing to say to one another—the less so when we consider that a dialogue in which the reply is separated from the question by a billion years is utterly irrelevant. Intelligent discussions might be held about which natural laws should be transformed and in what manner, but the time spent waiting for an answer would be too long for any effective action. The situation might be described like this: a certain ship, battling a storm, is so large that the machinist and navigator cannot co-ordinate their actions through a dialogue, since they must act too quickly for orders or replies. Every message is thus hopelessly late in relation to the actions that have already been initiated; as they arrive, the messages always refer to something no longer relevant. Similarly, communication in the universe occurs on the level of action, not in articulated messages. The civilizations do not fight, since it would do them no good; nor do they converse, since that would be meaningless. Gradually, over millions of years, their co-operation has become harmonized and synchronized. In the beginning, surely, confusions did arise when they misunderstood one another's creative work; traces of this can still be observed by astronomers. But that time is long past. Now, the exalted partners do their work wrapped in energetic silence, and realize their plans of cosmic stabilization or transformation so well that hardly any part of the primal universe that existed seven to eight billion years ago remains untouched. In the course of time, they transformed the entire universe in accordance with the strategy of the exalted civilizations, and everything within it—stars, dust clouds, galaxies, nebulae, as well as the laws directing them—originated in the game of this coalition. The evolution of matter is governed by collective reason, which is embodied in the multitude of the highest civilizations.

At first, this new cosmogony was roundly attacked, but it also gained adherents when it became clear that its hypotheses allow for deductions that agree with observed phenomena. For instance, the theory explains the expansion of the universe, since an expanding universe would be the most comfortable home for all the leading civilizations participating in the game at the same time. As soon as life appears on the planets of second-generation suns, followed by the flowering of intelligent societies, the psychozoic density (the frequency of the presence of intelligent beings) of the cosmos is altered as more and more civilizations occupy each unit of space. Should neighboring civilizations reach agreements, they might form coalitions whose activities could disturb the progress of cosmic processes. Therefore, in order to keep the psychozoic density of the cosmos constant, the ancient civilizations effected the continuous expansion of the universe. This at once explains why terrestrial astronomers have not been able to discover a single civilization near the sun: the distances between civilizations *must* remain great. Pulsars, those great, pulsing founts of radiation, are instruments used to synchronize and keep in phase the activity of still more gigantic, but imperceptible, systems determining the measurable aspects of space. And quasars, each of which is a cosmic furnace so mighty that its capacity exceeds the total radiation capacity of the Milky Way, are devices for radiating energy into space: or, rather, this is how they functioned several billions of years ago, since, due to their great distance from the earth, we see them in their distant past, when the exalted partners had just begun their co-operation.

The new cosmology also explains the fact, known since the 1960s, that mathematics takes many forms, and the form in which it developed historically on earth is only one of many possible varieties. The foundations of the universe were changed and reconstructed precisely through these different mathematics. The multitude of mathematical systems is an image of the multitude of possibilities of cosmogonic creation available at the dawn of cosmic history.

We will not dwell further on this new cosmogony of the twenty-first century, which considered the universe to be the result of intentional actions; nor will we describe at length the new philosophical syntheses based on its models. In brief, these posited a dialectical triad composed of a thesis: the universe created by God; its antithesis: the universe as a nonintentional object; and their synthesis: the empirical interrelation of the two previous models, which, developed further, does away with transcendence and replaces it with metagalactically plural reason. Let us instead point out the relations between the three variations of science fiction we have cited above and the canons of literature.

The first work—on preventing earthquakes—might be

best treated as an adventure novel or a technological "project" novel. What could be easier than to populate its setting with conventional literary characters. The second text—a hypothesis about human nature—might be written in a number of modes, from pseudorealistic to extravagantly grotesque. But whatever mode the author chooses, the central protagonists, whose personal histories are the individual embodiments of general phenomena, will conform to the canons of literature.

The third theme, however, does not seem conducive to the same kind of literary treatment—at least if the traditional narrative structures are upheld, complete with the characterization of heroes bound to specific situations. The intellectual adventures of the new cosmogony's creators will not be contained by traditional naturalistic or realistic narrative structures. We need not know about these scientists' wives, children, and acquaintances, any more than about Newton's or Planck's social and marital lives. Here we cannot simply present a social-cultural background against which purely personal events take place; the concept requires the chronicle of an idea, not of the vicissitudes of a few individuals. Of course, with sufficient obstinacy, we could apply the traditional solutions to this situation as well; but then all higher abstractions would be systematically detached from the minute human gestures, reactions, recognitions, personal conflicts, separations, etc., that make up the ordinary substance of the novel. Narrative structures of literature are incapable of synthesizing the "microscopic" elements of the cosmogonic scientists' everyday lives with the general hypotheses of their new cosmogony. To attempt such a synthesis would lead to a fractured work, with literary fragments, on the one hand, and discursive passages summarizing the new cosmogonic views, on the other. What is needed is an entirely new narrative structure, one that might be modeled on historiography, the biographies of scientists, or perhaps a collage of excerpts from scientific texts, press clippings, the addresses of Nobel laureates, or other facsimiles.

We are in no way suggesting that science fiction should renounce altogether its traditional structures of writing, in order to substitute as yet entirely unproven ones. Our example has a different purpose: it seeks to demonstrate that not all dramas and adventures of the human spirit in search of knowledge can be adequately represented through the traditional canons of the novel or epic narrative. In other words, the potential treasury of the narrative structures of science fiction has not yet been satisfactorily exploited.

In the course of writing this book, we suppressed the temptation to elaborate a metatheory of literary activity that would have dealt with the central structures of literature. Had we attempted it, we would have shattered the already too confining framework of this monograph. Our task here is not to elaborate a metatheory of science fiction, or even of literature in general, but to determine the extent of the field of paradigms, to define the boundaries that can contain the fruits of every possible creative production. Any metatheory of creative work must embrace every kind of cultural and intellectual effort that produces articulated and systematic products, such as music, novels, poems, buildings, sculptures, and philosophical systems. Such an inclusive theory cannot as yet be elaborated, but the time is approaching when it may be fully articulated. Its portents are visible in many fields. The philosophy of science, for example, strives for a metaperspective on the structure of scientific theories. There are analogous attempts in mathematics to define all those structures that are invariant elements of every branch of pure mathematics. Similar researches are being pursued in other fields, such as linguistics and anthropology. It would be most desirable, therefore, to extrapolate the system of universal constants shared by all researches operating with the technical term "structure." But this term does not mean the same thing to the linguist, the mathematician, the anthropologist, and the literary scholar—indeed, the latter two cannot even define it unambiguously. Now that structuralism has become a fashionable movement, its practice is full of abuses. But these should not be permitted to blind us to the generalizations—even though distant and difficult—to which this method can lead us. We suspect—although this has not yet been expressed as a definite hypothesis—that every kind of creative activity whose goal is constructive and whose starting point is an ensemble of elements and the rules for their treatment has the characteristics of a process unfolding in a theoretically closed configurational space, and that the topological qualities of this space are determined equally by the multitude of potential objects to be constructed and the unsurpassable limits of the space.

Structuralism, as a word and a method, had hardly been born when, in 1945, G. Evelyn Hutchinson—reviewing the anthropologist A. L. Kroeber's book *Configurations of Culture Growth,* in the third issue of *American Scientist*—proposed a hypothesis worth recalling. In his book, Kroeber compares the various periods of world history that were made memorable by extraordinary developments in philosophy, science, philology, sculpture, painting, drama, music, etc. In each of these cultural activities we can distinguish the embryonic and initial stages, when the parameters of all potentially realizable constructs that can be derived from the culturally accepted paradigms of artistic creation are already set—as yet, without anyone's being aware of it. (Where such newly accepted models originate is a different question altogether. But it is no more necessary for us to answer this question than for a biologist concerned with the evolution of organisms to answer the question, How did life originate on earth?)

In this early stage of the model, the whole stock of structures that can be derived from it is markedly indeterminate; in the course of their creative work, succeeding generations gradually determine the field of possible configurations, until it is completely delimited and exploited. The creative work of every historical period has developed within such limits or boundaries. Indirect evidence of this is provided by the following phenomena. Since the individuals more talented than the norm, whom we call geniuses, are the results of the rare intermingling of genotypes (the "winning numbers" in the "chromosome lottery"), and since these exceptional coincidences of genotypes are determined by the statistical regularities guiding population genetics, we might expect geniuses to be distributed uniformly along the axis of historical time. But in reality, this is not the case; the distribution of highly talented individuals is definitely not uniform and fortuitous. Hutchinson therefore constructs the following hypothesis based on Kroeber's insight: the chances of becoming an outstanding creator under different historical circumstances are not uniform, since an individual can create only within the field of paradigmatic structures that he finds prepared for him when he comes into the world. Those who are born in an early phase of the exploitation of a given "family of structures" face an enormously broad and difficult task, since the mass of virtual possibilities has not yet been defined. Those born in this state, writes Hutchinson, may perhaps gain the recognition of a small circle of enlightened cognoscenti, but they will not become as well known, or be able to found schools or movements quite as easily, as those who begin their work in the stage of *maximal* development of a given creative tradition. Thus, the sooner a genius is born within an artistic tradition, the more he can create; but if he arrives too early, he may go unnoticed, and, lacking "social reinforcement," may remain merely an unknown precursor. The person who arrives at the peak stage of a particular tradition can create a great deal, backed by strong social reinforcement. The artist who appears when most of the possibilities have already been exploited can, at best, become an original representative of a decaying tradition. Thus, the ascents and declines in cultural production—which are evident only in retrospect—are actually movements first toward, then away from the maximum of a certain curve. As a whole, this curve—which expresses the rate at which original productions derived from the embryonic stage proliferate—is additive, and is therefore only minimally dependent on individual successes. Thus, we do not consider 1616 the final date of Elizabethan drama as a whole, but as the year of Shakespeare's death.

Hutchinson attempts to formalize this process of slow growth, quickening, peaking, and decay in a logistical curve used in the study of demographic dynamics. The curve resembles a letter *S*, the center of which is characterized by exponential inclines, its beginning and ending by slight inclines (the Verhulst-Pearl curve). I will not cite the math with which Hutchinson supports his hypothesis, since it is, in any case, far too simple to grasp the phenomenon—as Hutchinson is well aware. The introduction of only two parameters—the quantity of still exploitable "degrees of freedom" and the quantity of already exploited degrees at time T, when a certain creator X appears in the world—cannot be adequate. The mass statistical (stochastic) nature of the process would certainly require a greater number of variables. But the greatest problem with any attempt to quantify creative phenomena is that, in order to verify the effectiveness of the formal apparatus, the works of certain artistic periods must be enumerated as elemental facts, a process that would unequivocally classify them in terms of their originality, and, coincidentally, in terms of their relative value—a hopeless task.

But we do have at our disposal a unique creative territory where the researcher is spared the problems caused by the subjective nature of any evaluation of cultural products, since, in this case, the evaluation has already been done for him, in an entirely unambiguous and incontestable way: this is the realm of natural development. Every emergent organism—such as the prototypes of "insect," "fish," "reptile," or "mammal"—is generated in a manner equivalent to those structural paradigms within which further evolutionary development "exploited every possible advantage for construction." In other words, the "insect" or "reptile" represents a certain stabilized "tradition" of creation within the field of variables determined by the given system's typology. The evolutionary process inexorably strives to exploit every possible organic combination inherent within the given prototype. (We need not puzzle over the quality of the various consecutive solutions that emerged in this way; evolution performs the evaluation of the product "in our stead," since only the "good" product survives.)

It is worth noting that not all of the prototypes mentioned above have developed an equal, or even a similar, number of variations in the course of their evolution. The structural paradigm of insects, for example, has proved to be an incredibly "fertile creative principle": at least as many species have developed from it as from all the other prototypes combined. Evolution also has its "creative stages," and its own "different schools and traditions." Its first triumphs were the creation of fishes, then it moved on to reptiles, while in our time it is primarily engaged in variations on the "theme" of mammals. Each type has, in its turn, gone through a slow initial stage of specialized differentiation, to arrive at a stage of the maximum richness of possible forms (consider but the variety of evolutionary reptile modifications), and finally to end up in an "age of epigonism" or repetitive fulfillment. Such a correspondence of natural and cultural creativity can hardly be an accident. The hypothesis immediately offers itself that there are, still unknown to us, higher regularities that determine the field of potential structures, regularities moreover that operate universally in the case of every type of material-information creation.

Recently, two English scientists, M. A. Ede and J. T. Law, used a computer to construct models of the embryonic formation process of certain organs, specifically the extremities. They fed the machine five series of data: four had to do with the behavior of embryonic cells, the fifth provided a model of the gene mechanism guiding their growth. In the course of its work, the computer produced only such models of extremity structures as had actually come about in the evolution of the different species of vertebrates, from fish to human. A slightly retrograde modification of the gene program resulted in the transformation of the structure of a leg into the structure of a fin. Neither machines nor programs exist for making models of complete organic systems; nonetheless, as the above example shows, we can already observe in experiments how the initial paradigm "determines" the field of possible constructions.

In order to continue the original creation, we must always introduce new elements into the paradigm's system. In literature (which is of greater interest to us here than evolution), there is a fairly universal intuition that every "great," "original" narrative model has already been discovered (and, moreover, quite long ago). But this is only a relative truth. While it is true, with respect to historically known conditions, that the narrative structures have been exhausted, it is equally true that civilization, by creating new problems, also provides new possibilities for literature. Our age, for example, is marked by the decline of conventional structures of ethical judgment all over the planet, since it is now within our power instrumentally to execute the Last Judgment. This fact is lost on the writer who would bring alien visitors to a destroyed earth and have them deliberate among the ruins about "which side was right." One can speak of one's rightness, i.e., as representing the correct path, only as long as someone is left behind to evaluate what has happened. It is meaningless to discuss either side's being right or wrong when total destruction has become possible; the only argument worth articulating on the verge of the ultimate catastrophe is that the catastrophe must be averted. At that point, truth, in the sense of "good" and "bad," is irrelevant as long as it is not bound up with the only program that has not lost its meaning, namely, the project of preserving humanity. The type of writer referred to above wishes to preserve the traditional standards of judgment of the pre-Atomic Age, but since he is writing in a pseudo-realistic style, he does not want to bring the Good Lord Himself down to the ash-covered earth—and so sends "aliens" instead, to continue the same argument that led to the cataclysm, on behalf of the no longer existing earthlings. This is a classic example of the helplessness of thought when it is shackled by inadequate narrative structures.

Every culture has codes that delineate the phenomena it considers "normal," and others that "deviate" from these and come into different degrees of conflict with the regulating norms. When phenomena that are predominantly associated with the stock of normal descriptive structures are depicted through discrepant structures, the result is often comic. Indeed, many humorous works owe their existence to this rule of displacement—they are intentionally "erroneous messages," cast in descriptive structures that are normally inappropriate to the phenomena described (for example, the description of a family quarrel in terms of natural phenomena, such as typhoons and volcanic eruptions; or, inversely, anthropomorphizing the volcano or storm). Generally, comic effects are generated when the use of alien descriptive structures does not quite amount to a fundamental transgression of the given culture's decrees and prohibitions. Still, the more the described phenomenon's structure is subject to codification (social ritualization), the stronger the effect; thus, a chase scene presented on speeded-up film is not as comic as a speeded-up funeral.

The simplest procedure, as we know, is simple inversion. What could be simpler than to achieve new effects by inverting the conventional structures petrified by tradition: this is the principle behind Mark Twain's "antistories." It would be easy to construct a theory of literature if only the descriptive language did not damage the reality depicted in the work, but simply represented it in a different mode. The problem is that this is not the case at all. Language, the instrument of description, is also the creator of what it describes. (Language can describe itself, as well, thereby becoming an object, and not only in a linguistic sense; for the language that describes language has a different semantic function from the language it describes.)

As cultural prohibitions weaken, it becomes impossible for literature to confront them. An approach that a century ago would have been considered "blasphemous" or immoral now rises to the level of artistic innovation. To cite the most readily available example: dispassionate descriptions of things that customarily are not permitted to be presented "coldly." This is the principle behind Kafka's "In the Penal Colony," and this is how modern experimental prose often describes the sexual act. The result is a kind of culture shock—characteristic of many works by Henry Miller *(Tropic of Cancer, Sexus, Nexus, Plexus),* in which the author describes his characters with great precision as though they were machines, naming their bodily parts by their functions, and intentionally ignoring all relevant social-erotic taboos. But in such situations we can at least distinguish between the structure of the description and the immanent structure of the object described. This becomes impossible when the reader has no knowledge whatsoever of the object and has none of the normative guidelines fixed in his memory for how the object "should be" described. These "wobbles" of perception cannot be enclosed within any strictly conceived semantic theory, because they are problems of practice, in which the receiver of the information is an inseparable part of the informational system. (For us, living in the present, the Man on the Moon can already be a specific person, or a specific historical event, whereas for people living only a few years in the past he was a purely fantastic creature, whose fictive nature deprived him of the solid objective qualities unique to intersubjectively demonstrable facts.)

The principle of "transposing and displacing" descriptive structures in relation to their objects can produce valuable results, both aesthetically and epistemologically. But when this "mix-up" is the result of a writer's ineptness and ignorance, the narrative clings to any available structure like some fragile vine, and the effort can only end in failure. This sort of indifference usually reduces anthropological problems to stereotypical adventure novels, social phenomena to psychological and personal phenomena (e.g., the conflict of two cultures played out as if it were the conflict of two individuals), and the alternations of cultural codes and norms to primitive reorientations on the order of "Aha! so this is how it should be done!" By the same token, escape from realistic dilemmas into illusory solutions is the general rule (for example, the resolution of the conflict between socialism and capitalism through the arrival on earth of "highly developed cosmic beings" who compel humans to live in peace, etc.).

The crisis of art in our age stems from the general disappearance of normative rules of action, which in turn results from the erosion of a view of culture as sacred and unquestionable because its commandments form a more ancient code than do civil laws. Whether one could break cultural rules if they became inconvenient was a question that one simply could not pose publicly in former times. It was empiricism that proved to be culture's Trojan Horse, since its principal criteria are those of utility, which naturally raise questions of comfort and convenience. For empiricism, the only inviolable barrier is the totality of the attributes of nature it calls the body of physical laws. Thus, observing the human world from an empirical standpoint necessarily leads to the complete relativization of cultural norms everywhere where they impose "un-founded" imperatives and restraints. Art can never be content with the basic stock of prohibitions that empiri-cism respects—merely because it cannot transgress them —for that would reduce art to nothingness. If art were to confine itself to the goals of empirical knowledge, it would begin to resemble empiricism more and more, until it became a faint replica, a shadow of science.

Art—and specifically literature—had in its province structures inherited from a venerable past governed by the untouchable norms of religious doctrines and myths. Lit-erature has all but completely exhausted these models, and it has not been enriched by new ones, for the sources of such structures have dried up. It is irrelevant whether they dried up when their creative power was naturally exhausted or whether the invasion of technogenic pragma-tism had dammed them before they could reach maximum potential. Even if such latent, historically untested mean-ing-structures might still be "inventable" in theory, they would be of no use for either humanity or art. A structure of significations that had never shone with the light of sacred solemnity, and had never been treated with the respect, fear, and love with which humans react to the presumed presence of the transcendental secret, would have no value for art.

The collapse of every kind of taboo created a freedom so vast that literature quickly began to feel acutely uncom-fortable. From there on, its only forum of appeal is cul-ture, in a necessarily nonsacred sense. Literature can still operate with the model structures generated by this secu-lar culture. But the sense that all the actual, synchronically functioning structures of the cultural field are unsatisfactory has led to hybridizing techniques, with combinations of extremely divergent structures and their superimposition over one another. For example: the deterministic structure of myth alloyed with the indeterminate structure of reality, as in Mann's *Doktor Faustus,* Joyce's *Ulysses,* or Frisch's *Homo Faber.* The principle of such works is allusion. The writer must arrange his ostensibly realistic material, drawn from the fund of common experiences, in such a way that its resemblances to the structure of some venerable myth (Faust, Odysseus, Oedipus) is evident to the reader. The reference to myth not only serves to give a lofty sanctification to things that would ordinarily be meaningless, however. Myth can also be parodized, treated iconoclastically, or even forcibly demolished. In *Lolita,* Nabokov discredits the myths of the innocence and angelic purity of adolescent girls, for in the novel it is the young girl who seduces the would-be ravisher, not he who defiles her. Elsewhere, as in Nabokov's more recent novel *(Ada, or Ardor),* the author exploits the cultural arsenal of prohibitions against incest in a "ludic" mode, by extending them to other relations parodistically superimposed over one another: between blood relations of a certain family, between the signs of the code invented by the incestuous lovers, between the "aristocracy" and the "plebeians." Even empirical truth contradicts the postulates of the incest taboo, because, as it turns out, due to the sterility of incestuous relationship, "nothing would have come of it anyway."

The hallmark of such creative strategies is their authors' constant search for ever more emphatically expressed resistance. As far as creative motivation is concerned, this situation differs radically from previous historical situations, because the artist who believed in the uniqueness of the norms guiding and regulating his creative activity naturally did not consider it his primary responsibility to attack them. These norms were "programmed" into him, having been perfectly internalized intellectually and emotionally, and he obeyed them with grace. As a consequence, originality—the personal irreproducibility of the work—manifested itself above all in the form, since the lofty canons of religious-cultural faith did not prescribe down to the last atom what forms works of art were to take.

This "search for resistance"—the initially clandestine erosion of existing norms—developed gradually in art. Historically, it predates the advent of technical civilization, since *Don Quixote* already introduces the (otherwise quite ambivalent) collision of the "myth of chivalry" with a prosaic, nonmythic reality. But as more and more norms disappear from social praxis, literature faces ever-growing difficulties. Its predicament is beginning to resemble that of a child who has discovered that his incredibly understanding parents will let him break with impunity all his toys, indeed everything in the house. The artist cannot create specific prohibitions for himself in order to attack them later in his work; the prohibitions must be real, and hence independent of the writer's choices. And since the relativization of cultural norms has not so far been able to disturb the given characteristics of human biology, that is where writers today seek the still perceptible points of resistance—which is why literature is preoccupied with the theme of sex. But such tactics are short-lived, an accelerating escalation sets in, and the "law of diminishing returns" goes into effect. A cultural taboo is too fundamental to be a pliable barrier; once it has collapsed, it can no longer serve as a wall for wall-shattering rams to knock down, piece by piece. Thus the removal of the administrative, and not culturally generated, censorship barriers has produced such a lightning-fast "pansexualization" of literature that an amusing competition has begun in the description of the most obscene scenes.

Writers require the resistance of matter as they require air. In literature it is particularly meaningless to storm gates that are standing wide open. When the solid foundations of the cultural norms began to crack, then crumble, literature tried to establish for itself a special autarchy and self-sufficiency, but this could never be complete. Through its aesthetic means, its concrete works, literature attempts to prove what is both logically and empirically unprovable. This is the source of those polyvalent and ambiguous structures that are susceptible to divergent interpretations. Kafka's *The Castle,* for example, can be read as a caricature of transcendence, a heaven maliciously dragged down to earth and mocked, or in precisely the opposite way, as the only image of transcendence accessible to a fallen humanity. In the first instance, the revelation is compromised; in the second, its earthly interpretation. Works like this do not expose those main junctures that could reveal their unambiguous ontological meanings: and the constant uncertainty this produces is the structural equivalent of the existential secret. The secret is neither explained nor given a secondary meaning. It simply remains—not merely as an enigmatic reference, but vividly displayed as a tangible presence, created by the palpable, irreducible indeterminacy of the work's own structure. This "rock solidity" of the secret produced by the cunning structuration of the work is one possible response to the destruction of cultural norms.

The other possibility is the approach mentioned earlier: superimposition of very different structures, some of which are harmoniously striving toward the same goal, while others are in dissonance and headed toward collision. The result is a peculiar feeling of *depth,* since it is not always possible to determine anew which structure is fundamental and which is relative, or, rather, which is the "absolute system of relations," and which are the variables whose values must be interpreted with reference to the system's standards. In neither case is the guiding principle of the work arbitrary. Just as a breeder does not act blindly, fortuitously, or chaotically when he sets out to develop a superior strain from the original animal or plant prototypes, so a writer also does not act fortuitously when he cross-breeds and combines complex narrative structures in new ways. This does not mean that such a writer makes only pragmatic "improvements." For, as a linguistic system, the literary work is often simultaneously homogeneous and disparate: it can be perfectly coherent on some levels, in certain constitutive structures, whereas on other levels it may even be internally contradictory. Moreover, it may keep some of its potential structures open— leaving, as it were, a way out for the work to transcend its own sphere, as is done in "The Monkey's Paw," for example. This short story proposes an "appended transaction":

we must either accept the existence of the "ghosts," and thus the hypothesis of transcendence, which makes the story a coherent whole, or we discard the hypothesis, and the work dissolves into a series of fortuitous, coincidental events. The acceptance of transcendence is the price one must pay for the work to be coherent.

The approach of the French antinovel has been a most interesting example of creative exploration. This exploration moved into extremely dangerous territory; for the novelists, instead of "cross-breeding" various kinds of order, reached a point where the paradigmatic forms of order and disorder collided. This is an altogether understandable approach when the author's guiding principle is the attempt to maximize the number of semantic levels in his work. Every message loses its clarity when it is damaged, either through collision and intermingling with another message, or because it is caught up in a flood of "pure noise." If we posit that the task of literature is not ever to give a definitive explanation of what it presents, and is therefore to affirm the autonomy of certain enigmas rather than to enter into explanations, then the most enigmatic of possible secrets is a purely random series. Every code that holds a hidden message has some key that will open it and decipher it, except pure chance, which is not a mask that can be ripped away, and thus will always resist every attempt at a definitive understanding. There is an unintentional trap in this situation, however. Every chance situation can be transformed into a nonrandom system if one employs adequate supplementary hypotheses. For instance, one can state that the Scandinavian peninsula resembles the outlines of a seal not because of chance geographical occurrences, but because of intentional actions (i.e., God Himself willed it so when He created heaven and earth). In accordance with Occam's razor, one can designate any state to be intentional through such extraneous hypotheses, even when there is not a trace of intentionality in them.

The majority of the works of the French antinovelists are the semantic equivalents of the story of the Emperor's new clothes, in that a certain kind of "semantic nudity" —a lack of intentionality induced by turning on the "noise generator"—is seen by its recipients as "new clothes," or as a new type of literary narrative that is, in its own way, intelligible.

The antinovelists have employed this generator on many different levels of creative work. Nothing explains the superimposition of the following structure in *The Erasers:* (1) the myth of Oedipus, (2) "time loops," (3) the detective story. If we must, we can deduce the detective story from the myth, or the time loops from the investigation. But we cannot explain *without contradiction* the meanings of the whole triadic structure, unless we invoke an elaborate edifice of additional tortuous and arbitrary interpretations. In *The Erasers,* the heterogeneous narrative structures were aligned by chance. In another of Robbe-Grillet's works, *La Maison de rendez-vous,* the principle of chance operates in the fragmentation and gradual recombination of the plot, and the method of fragmentation is also random. (Thus, in *The Erasers,* the chance generator operates on the fundamental level, the level of total structure, whereas in *La Maison de rendez-vous,* it operates on the subordinate ones.)

From the reader's point of view, Kafka's method (endowing the total structure of the work with a multidimensional "indeterminacy") and that of the antinovelists (depriving the work of clarity of meaning through quasi-accidental interventions to produce obscurity) have similar semantic results. The reader, in the activity of reading, reconstructs the work in a way that explains it, and invests its partially random qualities with order. The only problem is that such works, like the ink blots of a Rorschach test, have no "true" interpretations. This state of affairs favors the writer, since the broader a work's field of cultural references, the better it defends itself against devaluation. Any work that extends deep roots through its semantic references can serve to integrate the culture in which it originates. But the practical problem remains of convincing the reader to make the necessary interpretative effort, to integrate through hidden cultural references something that appears, at first sight, impossible to integrate, because it is a product of chance. Readers must be persuaded that there is a real need for their efforts, and the author is aided in this by zealous and resourceful critics who become the veritable coauthors of such emphatically indeterminate texts (which is probably why these texts hold such an attraction for many of them).

The difference between the works constituting multi-structured sandwiches and the works that represent "noise-damaged" messages is the same as that between the information of an authoritative palimpsest and a pseudo palimpsest in which the work of monks illustrating a manuscript is interlayered with that of houseflies making their own "corrections." It goes without saying that the use of the noise generator as a creative device is not a trick, since cultural consensus approves of chance in the creative process (this fact is self-evident in the fine arts, and can be seen in such extravagant methods as pitching fistfuls of paint at the canvas, or tracking shoe soles dipped in paint across it, and so forth). It is another matter that these works are essentially mechanisms for the creation of semantic mirages, even when they give an impression of semantic richness; to search for their inherent significance is akin to searching for the objective correlation between the delusions and nightmares of the hallucinating mind. If the organizing principle of the work is chance, it cannot also be intentional at the same time. This means that intentionality is displaced onto another, more inclusive level of the work, since the choice to use chance as a creative technique is itself not a product of chance at all. It is the result of calculation or conscious intention—a calculation sanctioned by game theory, which tells us that if one player makes a random move, his partners must also resort to a random strategy. And this is where the anti-novel defeats us, because as readers we cannot justify chance strategies of reception, because that would mean the dissolution of articulation. As a result, we are forced to strive for integration, to pretend that the work is a coherent whole.

Another factor in this context is that literature makes general use of the structures of indirect description or allusion. Whenever something is completely known—i.e., when it is fully rooted in a given culture—it can be understood or deduced from a single sign or allusion, and because of this, any given state of affairs can be made vividly present for the "culturally practiced" reader even through the remotest circumstantial reference. This indirect description is a method of structuring a work by "remote control" guidance of the reconstructive efforts of the reader's imagination, not only in space, but also in time. Every description of a situation taken from the repertoire of culturally known situations invokes the repertoire of possible issues appropriate for it, and these issues are what the reader will anticipate. Within the framework of this structured anticipation, he will make his decisions by following directions given in the text, even when they are few, or barely present. (For instance, we can speak of blatantly erotic situations, or others which have hidden erotic content held in constant suspension.) Indirect, "remote control" description is dictated either directly by cultural norms (as, for example, the prohibitions against expressly naming and presenting what the culture considers too drastic), or by the author's individual choices and thought. In the latter case the descriptive structures are generally full of gaps; they are either incomplete, or they are dim mirror images of completely different, unnamed, unarticulated, and thus merely intuited, structures. This displacement of description into circumscription can happen gradually, and, furthermore, either discontinuously or continuously. To transform circumscription into description, one can either move over into the as yet unnamed, untouched center of the problem by offending cultural prohibitions (by speaking plainly of things that had been previously forbidden, or by introducing obscene words into the vocabulary); or one can go in the opposite direction and attach to the culture's generalized structures of reference still other references, whose proper territories are even further from the concrete events.

Because of this systematic refusal to speak plainly, the reader begins to feel unsure whether he or she really understands what the description is concretely about, and this gives rise to the semantic wavering that characterizes the reception of contemporary poetry. (Not all poetry, certainly: there is randomly composed poetry, but we can discern a high level of "systematic indirection" in the finest poems of Grochowiak, for example.) All these approaches have a common origin: as the level of the reception's indeterminacy rises, the reader's own personal determinations begin to waver. In practice, it is often impossible to determine whether a given narrative structure is only very indirect and elliptical, but essentially homogeneous, or one deliberately damaged by "chance noise," or even perforated, softened, and bent by another, discordant structure. Furthermore, since one can also create multilayered structures, even the concrete quality of the described object or situation can be transformed beyond recognition and reshaped from one level of articulation to another. Thus, it is often impossible to determine categorically whether the basic structure of description is an image of order or of chaos. We cannot always distinguish the consequences of chance intrusions from the planned transformations of a particular creative design.

How do these methods and vocabulary bear upon science fiction?

In the first place, we consider the primary unsolved problem of science fiction the lack of a theoretical typology of its paradigmatic structures. Since writers of science fiction do not even recognize the existence of this problem, the structures they use most frequently are neither aesthetically nor epistemologically adequate for their chosen themes. An example of aesthetic inadequacy is the practice of authors who attempt to write mimetic (pseudorealistic) works, and yet model such phenomena as "contact with another civilization" or an invasion from outer space after the relationship between detective and criminal. (Two aliens in Hal Clement's *The Needle*—a criminal and a detective—"hide" in the bodies of two humans; and the detective, the "symbiont" on the boy who has become his "host," searches for the criminal "concealed" in the body of an unknown man.) This cognitive process results in antiempirical narrative. In the closed ecological system of Isaac Asimov's story "Strike-Breaker," the director of the planet's sewage system is essential for the life of all. He is unexpendable, yet precisely because of his low social position he is the object of general contempt. The basic structural assumption is obviously antiempirical. Those who were of low status long ago, such as butlers, maids, or housekeepers, are today worth their weight in gold—and the relations between the housekeeper and her "masters" have changed radically. Nowadays, the housekeeper is almost the ranking member of the family; she is indulged, her caprices are respected, her whims attended to. Therefore, if we simply extrapolate this transformation of social conditions, we can see that, in Asimov's society, the strike-breaker cannot be a man on whom the life of the whole community depends and still be treated as a pariah.

The choice of narrative structures can often be antiempirical even in works that otherwise pose interesting problems, such as *Flowers for Algernon*. The structure of such works, reminiscent of the curves of normal distribution (or the inverted letter V), originates with a certain type of tale. In the action of *Flowers for Algernon,* a retarded young man's intelligence at first expands to an extraordinary degree, but no sooner does he experience the joy of intellectual creation than he regresses back to idiocy with terrible speed. The work is interesting psychologically, but it poses the problem of "intelligence expansion" as a "rise and fall" paradigm—which is not very plausible, precisely because its origin is in fairy tales; but, more important, it prevents the author from examining the socio-cultural dimensions of his hypothesis about the artificial increase of intelligence. He requires his newly intelligent hero's sudden restupefaction for dramatic effect (for this curve of the action presents a personal drama, the tragedy of an individual's fall from the heights of wisdom, arduously and barely achieved, and at the same time it creates the closed structure of events that automatically shapes the whole progress of the action). This extremely simple model would not be adequate to show the consequences of intelligence augmentation for the whole culture; and yet these consequences would be well worth treating. A university professor is a universally respected figure with a high social status, both because it is hard to become a professor (certainly not everyone who wishes to can become one) and because such specialists have a very important role in the culture (they are the ones who pursue creative research, and educate the host of specialists that make up the foundation of civilization). If, however, the augmentation of intelligence permitted anyone to become a university professor, and if this happened to become the easiest and most desirable solution for everyone, then society would have to defend itself against the destructive consequences of the situation. All those who would still be willing to be drivers, sewer workers, builders, or milkmen —in spite of the fact that it is within their power to attain the highest level of creative intelligence—would have to be generously compensated. They would come to be surrounded by the halo of noble renunciation of their innate potential for development for the sake of the community. If such a novel were written as a grotesque, the sewer worker would be the admired, respected, outstanding personage, the lofty spirit, whereas the professor would be merely a mediocre little man, a tiny gear in the great mechanism. (There would be, of course, many other consequences of such a "geniusification" process that we cannot consider here. They can be deduced with the proper reasoning. But an arbitrarily chosen closed structure of events, such as the paradigm taken from fairy tales, is certainly not appropriate for the task.)

Thus, science fiction takes flight from the models and methods of reasoning we have sketched here to the rigid, simplistic structures derived from fairy tales and detective novels. Because of this, the system of narrative structures generally used is muddled, and is inadequate for the futurological thematics of science fiction. In their choice of narrative structures, most science-fiction writers fail to consider any criteria of empirical adequacy for and the best possible arrangement of the objects and situations they wish to describe. They try to conceal the "dubious origin" of such structures (the detective novel, romantic stories, fairy tales), which leads to the unintentionally grotesque style characteristic of most science fiction.

The second problem of science fiction is the unresolved relationship of the narrative to phenomena that are as yet not associated with appropriate descriptive structures, since they are the first of their kind. Meetings with such unknowns at first lead inexorably to semantic-descriptive paralysis. At such times, the greatest dilemmas that humanity has, over the centuries, conquered in the course of its "natural gnoseological evolution" surface all at once. I am thinking here of the problems of categorizing and articulating new phenomena—and thus of their inclusion in the established schemes of identification and recognition—all the decisions that together give a final definition of what, precisely, a new phenomenon is, what it means, how it can be described, what ethics it implies, and so forth. Judging from the popular output, science fiction is completely unaware that such problems exist, that they must be considered and consciously and concretely resolved. If the new phenomenon is of a qualitatively different scale—contact with "aliens" in outer space, for example—it is all but certain that the repertoire of received, ready concepts will not be able to accommodate it without considerable friction. In all likelihood, a cultural, perceptual, and perhaps even a social-ethical revolution will be necessary. Thus, instead of the assimilation of the new, we must imagine the reordering and even the destruction of fundamental concepts, the revaluation of truths that were previously indisputable, and so on. To refer such phenomena to slick, closed, and completely unambiguous structures we must simply consider a flaw. We can learn which structures and methods are the most appropriate from the history of science—by examining, for instance, the vicissitudes of physics, with its whole series of conceptual-categorial revolutions. (In this sense, the completely fantastic, one-hundred-percent *invented* history of the "new cosmogony" would still be true to reality, at least structurally, since the succeeding conceptual orders were "turned inside out" and reordered either precisely in this way or similarly in each science's actual course of development.)

Science fiction can thus learn from science as well as from other forms of literature, such as experimental prose. But it cannot learn through the kind of passive imitation characteristic of the English new wave of science fiction. Experimental literature, as we noted, introduces into the creative process different forms of "noise" (the chance generator), and the criteria for selecting structures created in this way are purely aesthetic. Science fiction should add to these another and separate set of criteria for cognitive adequacy. (Some equivalent to "noise"—the significant dispersion of opinions, or the contradiction arising simultaneously from the same sources—arises whenever a particular science confronts a new and unfamiliar phenomenon, and enters the phase of rapid conceptual reorganization. At the same time, this "noise" is never pure nonsense; science has not simply slipped into chaos.) Authors of science fiction must therefore draw upon the paradigmatics of transformations.

Clearheaded "internal" critics of science fiction have long been displeased with the genre for its flight from the real problems of civilization. But criticism must deal not only with the text's relations to the external world. It must evaluate not only the structure of the things described, but also the structure of the description itself. The former generally determines the choice of themes, whereas the latter determines the sum total of the rules governing the treatment of the material—and these rules are not automatically defined by the chosen theme.

Science fiction remains mired in a stage of theoretical self-reflection similar to the aggressive, extreme reductionism of neo-positivism ("every science, from biology to psychology, must be reduced to the language of physics!"). When asked whether such a reduction is practicable or not, the enthusiastic neo-positivists answer yes, their opponents no, and that usually puts an end to the argument. The neo-positivists, amazingly, have not recognized the simple fact that the reductionist program is based on a fallacy. They wish to posit a logical dichotomy by way of exclusion of the middle, whereas the historical nature of scientific understanding does not allow such a conceptualization. Biology and psychology certainly cannot be deduced from modern physics. At the same time, we cannot be sure that the physics of the future (which cannot, in principle, be reduced to the physics of the present, just as Einstein's model of the universe cannot be reduced to Newton's, nor the indeterminacy of quantum physics to Laplacian determinism) might not create transitional branches that will intersect with corresponding branches of the biology or psychology of the future (general systems theory). For that matter, we might attain the synthesis in yet another way, as the cyberneticists envision it: the synthesis would come about, not on the level of particular sciences, but on the next higher level of abstraction, with the discovery of the constants common to all the branch sciences.

Science fiction is reminiscent of neo-positivism's aggressive reductionism in that it acts as if the miserable repertoire of the detective story and the adventure novel were sufficient for structuring any phenomenon in the whole spectrum of the infinite universe, regardless of its time, place, and degree of complexity, and all the situations in which human civilization may ever find itself. Thus science fiction designates its problems (contact with aliens, the spirit in the machine, the instrumentalization of values, etc.), but it does not embody them in narrative structures.

In summary, it is clear that among the criticisms leveled against existing science fiction, the most important is this matter of opportunities systematically squandered. We must deem it a serious flaw of the genre that it has no independent, rational, and normative criticism that is neither destructive nor apologetic and that is committed not only to science fiction, but also to the more encompassing relations between culture and literature on which the fate of both depends. For this reason, my intention has not been so much to write the definitive monograph on science fiction, but, instead, to prepare the outline of a rational, internal critique.

Translated from the Hungarian
by Etelka de Laczay
and Istvan Csicsery-Ronay, Jr.

COSMOLOGY AND SCIENCE
FICTION

These remarks owe their existence to a suggestion of Dr. R. Mullen, of *Science-Fiction Studies,* who received a review copy of *Cosmology Now,* edited by Laurie John, but felt the book too peripheral to the journal's concerns for an ordinary review. The title, too, is Dr. Mullen's choice. Therefore my remarks are addressed to the readers of *Science-Fiction Studies,* and they were written in German because my English is insufficient for the task.

1. *Cosmology Now* was written by several British scientists for the BBC in 1973. The American edition, the one on hand, appeared in 1976. A reviewer both well versed in the subject and malicious could claim with some justifica-

tion that the book would be better called *Cosmology Yes-terday*. If the cosmos is the most durable of things, this durability doesn't extend to the science that deals with its exploration. Even the best cosmological reference works written some seven or eight years ago are today totally out of date. The three life-years that *Cosmology Now* has now had have seen much change in cosmology. Since I don't have to write a "regular review," I will list only the most important innovations. The age of the cosmos is today estimated to be some twenty billion years. The experi-ments of Weber, who claimed to have registered gravita-tional waves, have been discarded, since his apparatus was of insufficient sensitivity. The health of the "steady state" theory, which denies the evolution of the universe from a zero point, has deteriorated noticeably. Scientists are in-clined to award the palms of victory to the theory of the Big Bang. Moreover, many of the things described in *Cos-mology Now* have lost their former, beautiful simplicity. For instance, there is now a whole "family" of black holes. In addition to the ones postulated originally, which were supposed to be the final stage of a collapsing neutron star, there have been new ones—for instance, partially revers-ible black holes. These may not be assumed to be "gravity graves," invisible for all eternity. And there are, espe-cially, the black micro-holes. As the new theory of Ste-phen Hawking, of Cambridge, will have it, these are ob-jects with the diameter of a proton and mass of a mountain range. Quite a lot of them are said to have been created at the time of the Big Bang. I mention the theory of Hawking, first, because it introduces the method of quan-tum mechanics into the field of the general theory of relativity, and, second, because it implies consequences that cannot be overlooked and may change our whole outlook. Although there are so far no irrefutable (empirical) proofs for the existence of *any* black holes, we cannot imagine any possible technological utilization of the *big* black holes, whereas one may consider the micro-holes as energy sources that can surpass the annihilation of matter by several million times, the so far energetically most potent reaction. Such a micro-hole is supposed to contain the energy of several million hydrogen bombs. *Sapienti sat.* There are other important discoveries, but I cannot enlarge this short aside into a "regular book review." Therefore—*finis.*

In our times, scientific works grow old very fast. *The Internal Constitution of the Stars* by A. Eddington enthralled me when I read it forty years ago, and it is still a magnificent book, but it must be read now as (genuine!) science fiction, because nothing in it corresponds any more with our present knowledge. In my opinion the same may happen with *Cosmology Now:* please take this remark as an *hommage.* This volume will remain readable, indeed exciting, but very little of its aesthetically appealing, lucid simplicity in its development of the model of the universe will survive the changes to come. I say this as a dilettante and a heretic who knows more about the history of science than about cosmology. The first conquerors of new knowledge always find it easier to proclaim that "God may be subtle, but He is not malicious" because the biggest hurdles are discovered by the next generation of scientists. But it seems to me that one of the main theses of *Cosmology Now* will remain valid: that the universe is a continued explosion extended over a time of twenty billion years that appears as a majestic solidification only to the eyes of a transient being like man. The question whether we are living in a rhythmically pulsating universe or in a cosmos that will finally dissolve into vacuum still remains to be answered. The pendulum of mutually exclusive opinions goes on swinging.

2. Now then, what is the relationship between cosmology and science fiction? The facts are clear: both universes, that of the writers and that of the scientists, grow ever more apart. The estimations of the "density of cosmic civilization" show this most evidently. The scientists, even the founders of CETI (Contact with Extraterrestrial Intelligences), feel compelled to attribute ever smaller figures to the psychozoic density in the cosmos, because the accumulating negative results of the "sky listening" (for signals) force them to do so. Science fiction takes not the slightest notice of such changes. Therefore for science fiction one of the biggest riddles of contemporary cosmology, the *silentium universi,* doesn't exist at all. But it would be totally wrong to reduce the divergence of the two universes to only one parameter, the one mentioned. Science fiction started its escape from the real cosmos even before the question was formulated why the universe remains silent so stubbornly. This flight has by now evolved into a "steady state"; science fiction has encapsuled itself so much against the space of cosmology that it is unwilling to receive any signals; that is to say, any news from the field of science, with the exception of what manages to make the front pages of the newspapers (such as the tale of the black holes). This encapsulement took place when

the authors got hold of two fantastic, very convenient inventions: unlimited travel in time, and unlimited travel in space. Thanks to time travel and faster than light the cosmos has acquired such qualities as domesticate it in an exemplary manner for storytelling purposes; but at the same time it has lost its strange, icy sovereignty. Science fiction doesn't know of the cosmos of colliding galaxies, the invisible stars sucked in by the curvature of space, the pulsating magnetic fields. Nevertheless, there is in science fiction not a single one of the civilizations of the "third stage" postulated by CETI, the civilizations that are, thanks to their applied science of astral engineering, able to control stellar energies. As far as their content is concerned, most of the civilizations in science fiction correspond to the state predicted for earth in 2000 or 2300, although structurally they have remained arrested instead in the nineteenth century, with their colonizatory tactics of conquest and their strategies of war, whose magnification is due only to the principle of "Big Bertha," the German supergun that shelled Paris during the First World War. Science fiction has not the slightest idea what could be done with a power of the magnitude of a sun, if it isn't used exclusively for the destruction of inhabited planets. And in science fiction, cosmic civilizations have no intellectual culture at all, because a future-oriented movement that claims to probe into the farthest future, and makes its home in a realm of naïvely contaminated, amateurish ideas on "primitive slave societies," must be held totally lacking in credibility. Science-fiction criticism often talks of a "sense of wonder" that the field is supposed to generate, but upon close examination that "wonder" divulges its close relationship to the tricks of a stage magician. As popular fiction, science fiction must pose artificial problems and offer their easy solution. The astonishing results of contemporary cosmology, which border on paradox, are of no use to science-fiction writers, because they cannot be tucked into the narrow fixed frame of the artificial cosmos. Any comparison, including that with the stage magician, isn't quite exact; the magician doesn't aim at anything beyond the production of some tricks, whereas the self-imprisonment that is characteristic of science fiction has made it unable to describe real space any more.

To do justice to science fiction, which looks so shabby when compared to the background of cosmology, it is necessary to explain its dilemma further. The sins of individual authors have always been relatively small. The development of the totally false, domesticated universe was a gradual process of self-organization, and therefore all together are responsible for the final deformation—and nobody. Thanks to the first science-fiction invention, all occurrences in space have become easily reversible, but the authors who "just" want to shine with a new version of time travel have forgotten the larger context. It is particularly due to these unnoticed relationships that nature was softened in the cruelty of the irreversible flow of time that is its hallmark. In order that space might not be used as another cruelty to man, it was "short-circuited" by another invention, i.e., annihilated. The fact that a domestication of the cosmos has taken place, a diminution that whisked away those eternally silent abysses of which Pascal spoke with horror, is masked in science fiction by the blood that is so liberally spilled in its pages. But there we already have a humanized cruelty, for it is a cruelty that can be understood by man, and a cruelty that could finally even be judged from the viewpoint of ethics—granted that one could take this blood seriously at all. By looking at it this way, we come to understand what science fiction has done to the cosmos; for it makes no sense at all to look at the universe from the viewpoint of ethics. Therefore, the universe of science fiction is not only minuscule, simplified, and lukewarm, but also has been turned toward its inhabitants, and in this way it can be subjugated by them, losing thereby that indifference which causes man to project continually new enigmas to be solved and secrets to be lifted, in the vain hope to get *there* the answer to the question of his own meaning. In the universe of science fiction there is not the slightest chance that genuine myths and theologies might arise, because the thing itself is a bastard of myths gone to the dogs. The science fiction of today resembles a "graveyard of gravity," in which that subgenre of literature that promised the cosmos to mankind dreams away its defeat in onanistic delusions and chimeras—onanistic because they are anthropocentric. The task of the science-fiction author of today is as easy as that of the pornographer, and in the same way. Now that all the real stops to the satisfaction of their impulses have been pulled, they can have their fling. But with the stops has disappeared the indescribable richness that can be conveyed only by real life. Where anything comes easy, nothing can be of value. The most inflamed desire must finally end in miserable dullness. Once the credible, the real barriers have been blown up, the process of falsification must go on; artificial barriers must be erected, and in this manner the stuffed waxworks come about, the miserable ersatz that is supposed to be cosmic civilizations.

3. Why is it impossible to regain the universe that has been lost to science fiction? One could claim that the laws of the market do not permit it—that today no authors and publishers would dare to subject the readers to a cure of giving up that would equal the renunciation of easy solutions to fictitious problems. True, it must be admitted that not everything in science fiction is rotten in the same degree. After all, there was once the cosmogonic fantasy of a Stapledon. But Stapledon, as an isolated writer, was still able to view the universe of cosmology, and not the humanized universe of science fiction. It should be kept in mind here that "humanize" in this context doesn't mean to "make more humane"; we know that among the animals there are no sexual murderers, and a sexual murderer can hardly be called a humane being.

It must be admitted that the universe presents the "peak of indigestibility" for fiction writing in the whole field of our experience. For what can you do as an author with the central subjects of cosmology—with the singularities? A singularity is a place that exists in the continuum just as a stone exists here; but there our whole physics goes to pieces. The desperate struggles of the theoreticians, going on for several years now, have only the purpose of postponing this end of physics, its collapse, by yet one more theory. In fiction, however, things like that cannot be domesticated. What heroic characters, what plot can there be where no body, however strong or hard, could exist longer than a few fractions of a second? The space surrounding a neutron star cannot be passed closely in a spaceship even at parabolic velocity, because the gravity gradients in the human body increase without a chance that they might be stopped or screened, and human beings explode until only a red puddle is left, just like a heavenly body that is torn apart from tidal forces when passing through the Roche limit. Is there therefore no way out of this fatal dilemma: that one must be either silent about the cosmos or forced to distort it? Cosmology shows us a way out.

Just as one may look at the knowledge of yesterday as a fantastic speculation—as I said about the famous work of Eddington—so one may imagine a cosmogony of tomorrow, dissimilar to the current one, but nevertheless understandable, since cosmic processes are accessible to us to the degree that they can be focused by reason. But nothing is today so much held in contempt in science fiction as reason. In this regard a total harmony unites the authors with the readers. Obscenity is no longer indecent —the intellectual has taken its place in the pillory. Science-fiction fans should be discouraged from perusing *Cosmology Now,* unless they are willing to free their imagination from its imprisonment to discover in the brightness of real suns the true face of nature.

Translated from the German
by Franz Rottensteiner

TODOROV'S FANTASTIC THEORY

OF LITERATURE

Since structuralism in literary studies is largely of French origin, let this attempt to ruin its reputation have as its motto the words of a Frenchman, Pierre Bertaux:

> At one time it was hoped that the beginnings of a formalization of the humanities analogous [to that of the "diagonal" or "formalistic" sciences] could be expected from structuralism. Unfortunately it appears today that precisely the loudest advocates of structuralism have let it degenerate into a mythology —and not even a useful one. This chatter that is now

called structuralism has apparently dealt a mortal blow to that rudimentary scientific beginning.[1]

I fully agree with this verdict. However, inasmuch as it is difficult to expose in a single article the barrenness of a whole school of thought—one moreover that has spawned divergent tendencies, since here every author has his own "vision" of the subject—I will limit myself to dissecting Tzvetan Todorov's book *The Fantastic.*[2]

The history of the degeneration of a conceptual apparatus that originated in mathematical linguistics, after it was mechanically transplanted into the domain of metaliterature, has yet to be written. It will show how defenseless

1. Bertaux is a Germanist, and he published the article quoted, "Innovation als Prinzip," in German in the volume *Das 198. Jahrzehnt* (Christian Wegner Verlag, 1969). —SL. The passage given in German in Dr. Lem's original text (from which the first sentence has been reduced to the bracketed phrase in our translation) reads as follows: "Unter 'Diagonalwissenschaften' (um den Ausdruck von Roger Caillos aufzunehmen) verstehe ich ungefähr das, was man auch 'formalistische' Wissenschaften nennt, also Disziplinen, deren Gebiet sich quer durch die herkömmlichen Fächer der Realwissenschaften zieht. . . . Eine Zeitlang hat man hoffen können, der Ansatz zu einer ähnlichen Formalisierung der Humanwissenschaften sei vom Strukturalismus zu erwarten. Leider sieht es heute aus, als ob gerade die lautesten Vertreter des Strukturalismus ihn zu einer Mythologie hätten entarten lassen—und nicht einmal zu einer brauchbaren. Das Gerede, das jetzt den Namen Strukturalismus trägt, hat den ursprünglich in ihm enthaltenen wissenschaftlichen Ansatz wahrscheinlich tödlich getroffen." —Charles Nicol, R. D. Mullen, Darko Suvin.
2. Translated by Richard Howard (Cleveland/London: The Press of Case Western Reserve University, 1973) from *Introduction à la littérature fantastique* (Editions du Seuil, 1970). All quotations from Todorov are from the pages of this translation. —R. D. Mullen.

logical concepts become when they are torn out of contexts in which they were operationally justified, how easy it is, by parasitizing on science properly speaking, to bemuse humanists with pretentious claptrap, disguising one's actual powerlessness in a foreign field beneath a putatively unassailable logical precision. This will be a rather grim, but instructive, history of how unambiguous concepts turn into foggy ones, formal necessity into arbitrariness, syllogisms into paralogisms. It will, in short, deal with a retrograde trend in French critical thought, which, aiming at nothing less than logical infallibility in theory-building, transformed itself into an incorrigible dogmatism.

Structuralism was to be a remedy for the immaturity of the humanities as manifested in their lack of sovereign criteria for deciding the truth or falsehood of theoretical generalizations. The formal structures of linguistics are mathematical in origin, and are, indeed, numerous and diverse, corresponding to branches of both pure and classical mathematics ranging from probability and set theories to the theory of algorithms. The inadequacy of all these leads linguists to employ new models, e.g., from the theory of games, since this furnishes models of conflicts, and language is, at its higher, semantic levels, entangled in irreducible contradictions. These important tidings have, however, not yet reached those literary scholars who have taken over a small fraction of the arsenal of linguistics and endeavor to model literary works using conflict-free deductive structures of an uncommonly primitive type—as we shall demonstrate with the example of the Todorov book.

This author begins by disposing of some objections which arise in connection with constructing a theory of literary genres. Deriding the investigator who would, before proceeding to description of a genre, engage in endless reading of actual works, he asserts—appealing to the authority of Karl Popper—that for the maker of generalizations it suffices to be acquainted with a representative sample from the set of objects to be studied. Popper, wrongly invoked, is in no wise to blame, since representativeness of a sample in the natural sciences and in the arts are two quite different matters. Every normal tiger is representative for that species of cats, but there is no such thing as a "normal story." The "normalization" of tigers is effected by natural selection, so the taxonomist need not (indeed should not) evaluate these cats critically. But a student of literature who is in like fashion axiologically neutral is a blind man confronting a rainbow, for, whereas there do not exist any good organisms as distinguished from bad ones, there do exist good and worthless books. And in the event, Todorov's "sample," as displayed in his bibliography, is astonishing. Among its twenty-seven titles we find no Borges, no Verne, no Wells, nothing from modern fantasy, and all of science fiction is represented by two short stories; we get, instead, E. T. A. Hoffmann, Potocki, Balzac, Poe, Gogol, Kafka—and that is about all. In addition, there are two crime-story authors.

Todorov declares, further, that he will pass over problems of aesthetics altogether in silence, since these are beyond the present reach of his method.

Thirdly, he debates the relationship of the Species and its Specimen. In nature, he says, the occurrence of a mutation does not modify the species: knowing the species tiger, we can deduce from it the properties of each individual tiger. The feedback effect of mutations upon the species is so slow that it can be ignored. In art it is different: here every new work alters the species as it existed heretofore, and is a work of art just insofar as it departs from a specific model. Works which do not satisfy this condition belong to popular or mass literature, such as detective stories, slushy love stories, science fiction, etc. Agreeing thus far with Todorov, I see what is in store for his method as a result of this state of affairs: the more inferior and paradigmatically petrified the texts which it undertakes to anatomize, the more readily it will reveal structures. Todorov, not surprisingly, omits to draw this conclusion.

Further, he discusses the question of whether one should investigate genres that have arisen historically or those that are theoretically possible. The latter strike me as coming to the same thing as a history of mankind a priori, but since it is easier to formulate a foolish idea concisely than it is to refute it concisely, I will let this pass. I will however remark here that there is a difference between taxonomy in nature and in culture which structuralism overlooks. The naturalist's acts of classification, say of insects or of vertebrates, evoke no reaction on the part of that which is classified. A futurologist might say that Linnaean taxonomy is not subject to the Oedipus effect (Oedipus got into trouble by *reacting* to a diagnosis of his fate). On the other hand, the literary scholar's acts of classification are feedback-linked to that which is classified, i.e., the Oedipus effect manifests itself in literature. Not straightforwardly, to be sure. It is not the case that writers, upon reading a new theory of genres, run straight to their studios to refute it by means of their next books. The linkage is more roundabout. Sclerosis of paradigms, as a stiffening of intergeneric barriers, arouses authors to a reaction that expresses itself, among other ways, in the hybridization of genres and the attack on traditional norms. Theoreticians' labors are a catalyst that accelerates this process, since their generalizations make it easier for writers to grasp the *entire* space of creative activity, with its inherent limitations. Thus the student of genres who establishes their boundaries causes writers to rebel against them—he produces a feedback loop by the very act of classification. To describe limitations on creativity thus amounts to drawing up a self-defeating prognosis. What could be more tempting than to write what theory prohibits?

The constriction of the imagination that is inherent in a dogmatic mentality, such as is represented by the structuralist, manifests itself in the belief that what he has found to be barriers to creativity can never be transgressed by anyone. Perhaps there exist intransgressible structures of creativity, but structuralism has not come within reach of any such. Rather, what it proclaims to us as bounds of creativity is really quite an antique piece of furniture—to wit, the bed of Procrustes, as we shall show.

Coming to matters of substance, Todorov first of all demolishes past attempts at defining the fantastic. After crossing off the efforts of Northrop Frye, he lights into Roger Caillois, who had the bad luck to write that a "touchstone of the fantastic" is "the impression of irreducible strangeness" (p. 35). According to Caillois, jeers Todorov, a work's genre depends on the sang-froid of its reader: if he is frightened, then we have to do with the (uncanny) fantastic, but if he keeps his presence of mind, then the work must needs be reclassified from the standpoint of the theory of genres. We will speak in the proper place of how the scoffer has here left his own method exposed to attack.

Todorov distinguishes three aspects of the literary work: the *verbal,* the *syntactic,* and the *semantic,* making no secret of the fact that these were formerly known as *style, composition,* and *theme.* But their invariants have traditionally and mistakenly been sought "on the surface" of texts; Todorov declares that he will look for structures on a deep level, as abstract relations. Northrop Frye, suggests Todorov, might say that the forest and the sea form a manifestation of an elementary structure. Not so—these two phenomena manifest an abstract structure of the type of the relation between statics and dynamics. Here we first come upon the fruits of spurious methodological sophistication, that congenital trait of structuralism, for it is plain to see what our author is seeking: *oppositions* which come to light on a level of high abstraction. Now, this one is wide of the mark, because statics is not opposed to dynamics but is a special case of it, namely, a limiting case. This is a small matter, but a weighty problem lies behind it, since it is in *the same way* that Todorov constructs his integral structure for fantastic literature. This, by the structuralist's decree, consists of a one-dimensional axis, along which are situated subgenres that are mutually exclusive in a *logical* sense. This is portrayed by Todorov's diagram: "uncanny : fantastic-uncanny : fantastic-marvelous : marvelous" (p. 44).

What is the "fantastic"? It is, Todorov explains, the

hesitation of a being who knows only natural laws in the face of the supernatural. In other words, the fantastic character of a text resides in a transient and volatile state during the reading of it, one of *indecision* as to whether the narrative belongs to a natural or a supernatural order of things.

The "pure" uncanny amazes, shocks, terrifies, but does not give rise to indecision (of the kind we would call ontological). This is the place of the horror story, which presents occurrences that are frightful, extraordinary, but nevertheless rationally possible. This genre extends off the diagram to the left, merging into "ordinary" literature—as a transitional link, our theoretician mentions Dostoevsky.

The fantastic-uncanny already gives occasion to the vacillations that evoke the sense of the fantastic. This is a tale the events in which are, as its reader at first supposes, brought about by the intervention of the Supernatural. Its epilogue, however, furnishes a surprising rational explanation. (Here belongs, for example, the *Manuscrit trouvé a Saragosse.*)

The "fantastic-marvelous" work is just the other way around—it supplies in the end explanations of an extramundane, irrational order, as in Villiers de l'Isle-Adam's *Véra,* inasmuch as the conclusion of this story forces one to acknowledge that the dead woman really rose from the grave.

And finally the "pure" marvelous, which again does *not* give rise to any vacillations between mutually exclusive types of ontic systems, has all of four subdivisions: (a) the "hyperbolic marvelous," stemming from narrative extravagance, as in the voyages of Sinbad, where he speaks of serpents capable of swallowing elephants; (b) the "exotic marvelous": here, too, Sinbad serves Todorov's purpose, when he says that the Roc had legs like oak trees—this is not a zoological absurdity, since to long-ago readers such an avian form may have seemed "possible"; (c) the "instrumental marvelous"—the instruments are fabulous objects such as the lamp or the ring of Aladdin; and (d) the "scientific marvelous," i.e., science fiction. Of this last subdivision, he says: "These narratives, starting from irrational premises, link the 'facts' they contain in a perfectly logical manner" (pp. 56–57). And: "The initial data are supernatural: robots, extraterrestrial beings, the whole interplanetary context" (p. 172). And: "Here the supernatural is explained in a rational manner, but according to laws that contemporary science does not acknowledge" (p. 56).

The scientific bibliography of the theory of "robots" forms a thick volume; there exists a world-renowned organization of astrophysicists (CETI) concerned with searching for signals emitted by Todorov's "supernatural beings," i.e., by extraterrestrial creatures; for our theoretician even the "interplanetary background" possesses supernatural properties. Let us, however, regard all these qualifications as slips of the pen. We may as well do so, since Todorov's theory would be fine if it contained only such defects.

As we know, Todorov calls the fantastic a transitional boundary state on an axis whose opposite extremes signify the rational system of nature and the irrational order of marvels. For a work to manifest its fantastic character, it must be read literally, from the standpoint of naïve realism, thus neither poetically nor allegorically. These two categories, according to Todorov, exclude one another with logical necessity, hence fantastic poetry or fantastic allegory is *always* impossible. This second categorical axis is perpendicular to the first. Let us clarify these relationships on a "microexample" of our own, given by a single simple sentence. The sentence "A black cloud swallowed the sun" can be taken, first of all, as a poetic metaphor (a thoroughly trite one, but that is beside the point). The cloud, we know, was only *figuratively* compared to a being capable of devouring the sun, since in fact it merely hid it from view.

Furthermore, it is possible, by dint of contextual suggestions, to substitute for the cloud, say, falsehood, and for the sun, truth. The sentence becomes an allegory: it says that falsehood may obscure truth. Again, this is a platitude, but the relations that hold are clearly apparent, and that is what we are after.

Now if instead we take the sentence literally, some uncertainties emerge that make it possible for *indecision* and, by the same token, the fantastic to result. The cloud, we know, "actually swallowed the sun"—but in what order of events, the *natural* or the *marvelous*? If it gulped it down as a fairy-tale dragon might, then we find ourselves in a fairy tale, in the "pure marvelous." But if it engulfed the sun as did a certain cosmic cloud in the novel *The Black Cloud* by the astrophysicist Fred Hoyle, we shift to science fiction. In this novel the cloud is made of cosmic dust, it is a "cybernetic organism" and it engulfed the sun because it feeds on stellar radiation. The explanation acquires rationality as a hypothetical extrapolation from such disciplines as the theory of self-organizing systems, the theory of evolution, etc.

To be sure, the results of our classification do not coincide with Todorov's, since for him science fiction is irrationalism embodied in pseudoscience. There is no point to arguing about Hoyle's Black Cloud. It is enough to note that science fiction is nourished by scientific revelations—e.g., in the aftermath of the heart transplants there appeared swarms of fictional works that described criminal gangs snatching hearts from the breasts of young people on behalf of rich oldsters. Even if this is improbable, it assuredly does not belong to any supernatural order of things. But after all, arbitration might reconcile the conflicting viewpoints by effecting, say, within the scope of Todorov's axis, a translation of some titles, at least, toward the pole of the Rational.

Things get worse when it comes to subgenres of the fantastic for which there is no place at all on Todorov's axis. To what genre should Borges's "Three Versions of Judas" be assigned? In this work Borges invented the fictional heresy of a Scandinavian theologian, according to which Judas, not Jesus, was the true Redeemer. This is not a "marvelous" tale—no more than any genuine heresy such as the Manichaean or the Pelagian. It is not an apocryphon, for an apocryphon pretends to be an authentic original, while Borges's text does not try to conceal its literary nature. It is not an allegory, nor is it poetry, but, since nobody ever proclaimed such an apostasy, the matter cannot be placed in the order of real events. Quite obviously we have to do here with an imaginary heresy, that is, with *fantastic theology.*

Let us generalize this interesting case. Let us recognize unprovable propositions, such as metaphysical, religious, or ontological assertions, as forming an "actual religious credo," a confession of faith, the affirmation of a world view, if they have entered in just this guise into the repository of the historic civilizations. From an immanent standpoint it cannot be discerned from any such proposition, whether it was uttered with the conviction that things are really as it claims, or whether it was enunciated nonseriously (in "ludic" fashion, thus nonassertively). If no philosopher named Schopenhauer had ever existed and if Borges had invented in a story a doctrine called "The World as Will," we would accept this as a bit of *fiction*, not of the history of philosophy. But of what kind of fiction, indeed? Of *fantastic philosophy*, because it was published nonassertively. Here is a literature of imaginary ideas, of fictional basic values, of other civilizations—in a word, the fantasy of the "abstract."

On Todorov's axis there is likewise no place for fantastic history, which did not happen but might have. This is a matter of so-called political fiction, telling what might have been if Japan rather than the United States had fabricated the atomic bomb, if the Germans had won the Second World War, and the like. These are not uncanny tales—at any rate, no more so than what has actually happened in the present century—and they are not marvelous, since it would hardly have taken a miracle to make Japanese physicists go to work building reactors, and also there is no question of the reader's being unsure about whether the narrated events are rational or irrational. And yet in just this way objective worlds are constructed, the nonexistence of which in past, present, or future is an irrefragable certainty. So what sort of books are these? Beyond a doubt, ones that fabricate a *fantastic universal history.*

Thus our Procrustes has not made place on his meager axis even for *actually* existing varieties of the fantastic— let alone "theoretically possible" kinds, for which there is a fortiori no room in his bed of torture.

Let us now take a closer look at Todorov's axis. It is of logical ancestry. The structuralist is indebted to the linguists, and they in turn adopted this simplest *structure of exclusion* from set theory, in that here the principle of the excluded middle holds: an element either belongs to a set or it does not, and forty-five percent membership in a set is impossible. Todorov ascribes to this axis a fundamental, because definitional, significance on the highest level of abstraction. However, the essential thing is not the axis but the reader's act of decision. Reading a literary work indeed calls for decisions—in fact, not just one, but an ordered set of them, which result in the genre classification of the text. The reader's decisions do not oscillate in only one dimension. Assuming as a working hypothesis that these are always decisions with respect to simple (binary) alternatives, thus dichotomous, one can enumerate such additional axes as:

(a) Earnestness : irony. Irony is calling a statement in question, either its linguistic level (this has been done stylistically by Gombrowicz) or its objective level. As a rule irony is in some measure reflexive. But lest the "deflation" of the utterance should become self-destruction on its part, this tactic stabilizes the reader's hesitancy, or renders futile the attempt at a definitive diagnosis with respect to the designated opposition. It achieves its optimum durability when the separation of an "ironic component" from a "serious component" in the text is not feasible. "Three Versions of Judas" is of just this kind.

(b) Autonomous (reflexive) text : relative text (referred to something outside itself). Todorov's "allegory" is a bag into which countless heterogeneous matters are stuffed. Culturally local (ethnocentric) allegory is something different from universal allegory. What is allegorical in the author's cultural sphere may be "mere entertainment" or "pure fantasy" for ethnically alien readers, in line with the saying: "Wer den Dichter will verstehen, muss in Dichters Lande gehen" ("Whoever wants to understand the poet must go to the poet's country"). The symbolism peculiar to Japanese prose may be unrecognizable by us, for precisely this reason. And again, symbolic character of a text does not necessarily make it allegorical. Whatever is a normative symbol (pertaining to taboo, say) of a given culture is by that very fact neither arbitrary, nor fantastic, nor "imaginary" for that culture's members. Whether a given text is autonomous or relative is determined by the community of culture between the author and his readers.

(c) Text as cryptogram : text as literal message. This is a variant of the foregoing opposition. The difference between the two is that in (b) it is a matter of relations among objects (events), but in (c) one of (linguistic) relations among utterances. Allegory is a sort of generalization signaled by *events-objects* (a man, as by Kafka, turns into an insect). The content of a cryptogram, on the other hand, can be anything, e.g., another cryptogram. From the fact that cryptograms exist it does not follow that everything is a cryptogram. From the fact that in certain cultures a part is played by themes concealed under relationships (social, familial) it does not follow that in every culture its relational character (its structure) must be a camouflage for meanings concealed in this fashion. This is why one feels a cognitive disappointment in reading Lévi-Strauss, because one cannot discover any reason, psychological, social, or logical, responsible for some meanings' functioning in the community in overt relationships (i.e., ones publicly called by their names), whereas others are "hidden" in the network of occurring relations and have to be reconstructed by abstraction. Here for ethnological structuralism there lies in wait the same bottomless pitfall that menaces psychoanalysis, since as in psychoanalysis it is possible to impute to the analysand's every word the status of a "mask" concealing *another*, deeper content, so in structuralism it is always possible to hold that what occurs as relations in a culture is inconclusive and unimportant, because it represents a "camouflage" for *other* concepts, those that will only be brought to light by the abstract model. Neither of these hypotheses can be verified, so they are nonempirical with respect both to assumptions and to methods.

One could go on enumerating such oppositions. Superimposing their axes, so that they form a multidimensional "compass card"—i.e., a co-ordinate system with multiple axes—we obtain a formal model of the situation of the reader who has to make repeated decisions about a complexly structured text. Not all texts activate the decision process along all the possible axes, but a theory of genres must take into account at least that class of decisions which cumulatively determines the genre classification of what is read.

It should be emphasized that particular decisions, until they are made, are dependent variables. Once we have concluded, for example, that a text really *is* ironic, we have thereby altered the probabilities of specific decisions on other axes.

The perfidy of modern creative writing lies just in making life—that is, semantic decisions—difficult for the reader. Such writing was emphatically initiated by Kafka. Todorov, unable to cope with Kafka's texts by means of his axis, has made a virtue of methodological paralysis, taking his own perplexity out into the deep waters of hermeneutics. According to him, Kafka conferred "complete autonomy" on his text; he cut it off from the world in all directions. The text seems to be allegorical but is not, since there is no way of ascertaining to what court it addresses its appeal. Hence it is neither allegorical nor poetic nor realistic, and if it can be called "fantastic," then only in the sense that "dream logic" has engulfed the narrative together with the reader. ("Son monde tout entière obéit à une logique onirique sinon cauchemardesque, qui n'a plus rien à voir avec le réel" [p. 181].) *Ita dixit* Todorov, without noticing that he has hereby abandoned all his structuralizing.

Todorov's conception of Kafka's works as totally lacking an address (as reflexive) in the real world ("n'a plus rien à voir avec le réel") has become popular also outside structuralist circles, I think, as a result of intellectual laziness. These works, boundlessly veiled in meanings, seem to signify so much at once that no one knows what they mean *concretely*. Well, then, let it be that they simply mean *nothing*, whether referentially, allusively, or evocatively.

If there existed an experimental science of literature concerned with studying readers' reactions to deliberately prepared texts, it would prove in short order that a text wholly severed from the world with regard to its meanings can be of no interest to anyone. References of expressions to extralinguistic states of affairs form a continuous spectrum, ranging from ostensive denotation to an aura of allusions hard to define, just as recall of things seen to our visual memory ranges from sharp perception in broad daylight to the vagueness of a nocturnal phantom in the dark. Consequently, a boundary between "undisguised reference" and "hermetic autonomy" of a text can be drawn only arbitrarily, because the distinction is extremely fuzzy.

A representative of impressionistic criticism might say that Kafka's writing "shimmers with mirages of infinite meanings," but an advocate of scientific criticism must uncover the tactics that bring this state of things about, not hand the texts a charter certifying their independence of the visible world. We have sketched above a way of effecting the transition from texts that are decisionally unimodal, simple ones, such as the detective story, to those that are n-modal. A work that embodies the relational paradigmatics of the "compass card" thereby sets up an undecidability about its own meaning in that it persistently defies that "instrument of semantic diagnosis" which every human head contains. There then takes place the stabilization of a shaky equilibrium at the crossroads formed by the text itself, since we cannot even say whether it is definitely in earnest or definitely ironic, whether it belongs to the one world or to the other, whether it elevates our vale of tears to the level of transcendence (as some critics said about Kafka's *The Castle*) or whether, on the contrary, it degrades the beyond to the temporal plane (as others said about *The Castle*), whether it is a parable with a moral expressed by symbols from the unconscious (this is the thesis of psychoanalytic criticism), or whether it constitutes "the fantastic without limits"—which last is the dodge our structuralist uses.

It is strange that no one is willing to admit the fact of the matter: that the work brings into head-on collision a swarm of conflicting interpretations, each of which can be defended on its own grounds. If what we had before us were a logical calculus, the sum of these conflicting judgments would clearly be zero, since contradictory propositions cancel one another out. But the work is just not a logical treatise, and therefore it becomes for us, in its semantic undecidability, a fascinating riddle. "Single-axis" structuralism fails utterly for it, but the mechanism of undamped oscillation of the reader's surmises can be formalized by a topology of multiple decision-making, which in the limit turns the compass card into a surface representing continuous aberrations of the receiver. However, the structuralist model even as we have thus amended it is not fully adequate to a work such as Kafka's. It falls short because its axiomatic assumption of disjointedness of opposed categories (allegory : poetry, irony : earnestness, natural : supernatural) is altogether false. The crux lies in the fact that the work can be placed on the natural and the supernatural level *at the same time,* that it can be at once earnest and ironic, and fantastic, poetic, and allegorical as well. The "at the same time" predicated here implies contradictions—but what can you do, if such a text is founded just on contradictions? This is made plain by the throng of equally justified but antagonistic interpretations that battle vainly for supremacy, i.e., for uniqueness. It is only mathematics and logic and—following their example—mathematical linguistics that fear contradictions as the Devil fears holy water. Only these can do nothing constructive with contradictions, which put an end to all rational cognition. What is involved is a trap disastrous for epistemology, in that it is an expression that contradicts itself (much like the classic paradox of the Liar). Yet literature manages to thrive on paradoxes, if only on ones strategically placed—precisely these constitute its perfidious advantage! Not, to be sure, from its own resources. It has not invented such horrendous powers for itself. We find logical contradictions ready-made, firstly in culture: for—to take the first example to hand—according to the canons of Christianity, whatever happens happens naturally, and at the same time it happens by the will of God, since nothing can be apart from this. The nontemporal order thus coexists with the temporal—eternity is in every moment and in every inch. The collisions of behavior provoked by this "overlapping" predication are buffered by successive interpretations of dogma, e.g., in a species of theological consent to the use of anesthesia in childbirth. Nonetheless there is a contradiction involved that culminates in "Credo, quia absurdum est." Secondly, overlapping categorizations of percepts become the norm in dreams as well as in hyponoic states, thus not only in psychiatric symptomatology (cf. Ernst Kretschmer, *Medizinische Psychologie*). The coexistence in apperception of states of affairs that exclude one another both empirically and logically is, consequently, a double regularity—cultural and psychological—on which structuralism finally breaks every bone in all its "axes." Thus the whole literary-critical procrustics or catalogue of adulterations, errors, and oversimplifications formed by this *Introduction à la littérature fantastique* is of value only as an object lesson illustrating the downfall of a precise conceptual apparatus outside its proper domain.

We still have with us the dilemma of the hard-headed reader, who, if he is not scared by a ghost story, relabels it with respect to genre. Todorov would hold such a receiver to be an ignoramus who ought to keep his hands off literature. But when we examine the situation in which someone reads an "uncanny" or a "tragic" text and splits his sides laughing, we will realize that this situation can be explained in either of two ways. Perhaps the reader is in fact a primitive oaf who is too *immature* to appreciate the work, and that is an end of the problem. Or perhaps the work is kitsch and he who laughs at it is an experienced connoisseur of literature, so that he cannot take seriously what the work presents as serious, i.e., he has *outgrown* the work. In the second case the text really does change its genre: from a story about spirits (intentionally uncanny) or about galactic monarchs (intentionally science-fictional) or about life in high society (intentionally edifying romance) it turns into an unintentional humoresque.

Todorov bars saying anything at all about an author's intentions—to mention these amounts to covering oneself with the disgrace of "fallacia intentionalis." Structuralism is supposed to investigate texts only in their *immanence*. But if one is free to recognize, as Todorov does, that a text implies a reader (not as a concrete person but as a standard of reception), then in accord with a rule of symmetry one should recognize that it also implies an author. Both of these concepts are indissolubly connected with the category of messages, since a message, in information theory, must have a sender and a receiver.

The words of Roger Caillois about "the irreducible impression of strangeness" as a touchstone for the fantastic represent the psychological correlate of the linguistic state of things constituted by the full-valued character of the artistic text, which guarantees that it is not kitsch. The irreducibility of the impression certifies the authentic values of the text and thereby abolishes the relativism typical for writing with unwarranted pretensions, which produces kitsch as an incongruity between intention and realization.

The relativism of kitsch lies in the fact that it is not kitsch for all readers, and, what is more, it cannot be recognized as kitsch by those who esteem it. Kitsch identified as such forms a special case of paradox within the set of literary works; namely, contradiction between the reactions anticipated by the text and the reactions that its reading actually evokes. For the uncanny is incompatible with nonsense, physics with magic, the sociology of the aristocracy with the scullery's notions about it, and the process of cognition with the adventures of puppets called scientists. Thus kitsch is a product counterfeited to pass for what it is not. The contradictions in interpretation of Kafka's writings not only can but must be grasped by the reader; only so, thanks to "indecision of manifold scope," will he apprehend the aura of mystery established by the text. Per contra, the contradiction specific to kitsch must remain unrecognized by its readers, since otherwise generic disqualification of what has been read will take place. The reading of kitsch *as kitsch* is nonimmanent—the reader appeals to his own superior knowledge about how a work of the given kind ought to look, and the chasm separating what ought to be from what in fact is amuses him (or offends him).

Because our superior knowledge decreases as the themes of literature become increasingly remote from reality, kitsch takes up residence in regions inaccessible to the reader: in the palace, in the far future, among the stars, in history, in exotic lands. Every literary genre has its masterwork-ceiling, and kitsch, by a tactics of crude mimicry, pretends to have soared to such an altitude. Todorov, fettered by the immanence of his procedures, has deprived himself of any possibility of recognizing mimicry of values, and accordingly his implicit reader *must,* by dint of solemn exertions, see to it that the silliest twaddle about spirits sends chills up and down his spine. On pain of a structuralist curse he is forbidden to poke fun at such rubbish; since structuralism establishes absolute equality in literature, the right of citizenship that the text usurps for itself is a sacred thing.

A possible rejoinder at this point would be that idiotic stories are written for idiotic readers. And indeed, we observe this state of affairs in the book market, dominated by the laws of supply and demand. But this is not an extenuating circumstance for a theory of literature. A "theory" is synonymous with a generalization that applies without exception to all elements of the set under investigation. Since the structuralists' generalizations balk at applying thus, or, more precisely, because when they are made to apply thus everywhere they yield such nonsense as no advocate of the school would like to acknowledge (for structural equivalence democratically places the counterfeit on an equal footing with the masterpiece), the theoreticians carry out certain sleight-of-hand manipulations when they assemble their materials for public dissection. They place on their operating table, to wit, only what has already earned a respectable reputation in the history of literature, and they conjure away under the table works that are structurally of the same kinds but artistically trashy. They have to proceed thus, because their method impels them toward simple texts such as the detective story; their overweening ambitions, on the other hand, toward celebrated works. (Kitsch, being subject to relativization in the process of reception, is not the structurally simplest case, for it *seeks* to be one thing and *is* in fact another; the detective story, on the other hand, devoid of pretensions, is decisionally unimodal.)

Now we can more readily understand the make-up of Todorov's bibliography, as to the names (Balzac, Poe, Gogol, Hoffmann, Kafka) and the works it includes. The theoretician has taken as his "sample" that which could not involve him in difficulties, since it had already passed its cultural screening examination and by that token could give him no trouble. A therapist, if he were to proceed analogously, would take as patients only robust convalescents. A physicist would test his theory only on facts that he knew beforehand would confirm it, carefully avoiding all others. Let us spare the structuralist the description that the philosophy of science would give to such a method of selecting "representative samples."

A theory of literature either embraces all works or it is no theory. A theory of works weeded out in advance by means beyond its compass constitutes not generalization but its contrary, that is, particularization. One cannot when theorizing discriminate beforehand against a certain group of works—i.e., not bring them under the scope of analysis at all. A taxonomically oriented theory can set up a hierarchy in its subject matter—i.e., assign nonuniform values to the elements of the entire set under investigation —but it should do this openly, not on the sly, and throughout its whole domain, showing what sort of criteria it employs for making distinctions and how they perform their tasks of evaluation.

These obligations are binding not for humanistic studies alone. They stem from the set of directives to which all scientific cognition is subject. A zoologist cannot ignore cockroaches because they're such nasty little beasties, nor a cosmologist ignore the energy balance of quasars because it makes his calculations blow up in his face. The sleight-of-hand artist's activities are not always and everywhere admirable. So, we conclude, if structuralism desires to avoid expulsion from among the sciences, it must rebuild itself completely, from the ground up, since in its present state it is, in the words of Pierre Bertaux, a procedure that from its point of departure in logic has strayed into useless mythology.

Translated from the Polish
by Robert Abernathy

UNITAS OPPOSITORUM: THE PROSE OF JORGE LUIS BORGES

I admit that this essay is a very subjective review of Borges's fiction. If someone asked me why I am stressing the subjective aspect of this piece of criticism, I would be hard-pressed to give a conclusive answer. Perhaps because I have been trying for years to enter the territory in which the Argentinian's best work was created, although I went by quite another road. Therefore his work is very close to me. At the same time it is foreign to me, for I know from my own experience the traps into which he has sometimes fallen in his writing, and I cannot always approve of his literary methods.

Nothing could be simpler than to list Borges's best

stories. These are: "Tlön, Uqbar, Orbis Tertius," "Pierre Menard—Author of the Quixote," "The Lottery in Babylon," and "Three Versions of Judas."

I justify my preference in the following way: each of the stories mentioned has a double-decker, perverse, but logically perfect structure. Viewed superficially, they are fictionalized paradoxes of the Greek type (Zeno's, for instance[1]).

In "Tlön, Uqbar, Orbis Tertius," Borges bases the story on the idea of reversing our concepts of "idea" and "reality." Borges suggests that a secret society has created a new world where the mind creates its own external objects, and the only external objects are those created by the mind.

In "The Lottery in Babylon" Borges contrasts two mutually exclusive explanations of the universe: (statistical) chance, and (total) determinism. Usually we consider these notions incompatible. Borges tells of a world system based upon a lottery, and reconciles two cosmological explanations without destroying the logical bases of each system.

"Pierre Menard—Author of the Quixote," on the other hand, is a satire on the uniqueness of the act of artistic creation, logically driven to its utmost point. (In this story Pierre Menard seeks to rewrite "Don Quixote" precisely —without copying it. The story shows the paradoxes behind the idea that art is created necessarily and uniquely. Borges reduces the idea *ad absurdum.*)

---

1. The difference is that Zeno's paradoxes confront the trivial interpretation of physical processes with the contradictory results of their purely logical interpretation, whereas the paradoxes of Borges are directed toward the universe of *cultural* facts.

Finally, "Three Versions of Judas" is a logically un-provable heresy.[2] Borges builds a fictitiously heterodox system of Christian dogmatics in which he "proves" that Judas not Jesus was the Christ. In its "radicalism" this fictitious heresy surpasses all historical types of heresy.

In each story we can find the same kind of method: Borges transforms a firmly established part of some cultural system by means of the terms of the system itself. In the fields of religious belief, in ontology, in literary theory, the author "continues" what mankind has "only begun to make." Using this *tour d'adresse* Borges makes comical and absurd those things which we revere because of their current cultural value.

But when we look at Borges's work only superficially we see the "comical-logical" effect alone. However, each of these tales has in addition another—wholly serious—hidden meaning. At base, his curious fantasy is, I claim, quite realistic. Only after some thought do you first note that the heterodoxy contained within "Judas," for in-

---

2. Strictly speaking, what has been said isn't true insofar as there are no systems of belief (of either an orthodox or a heterodox nature) that would not hide contradictions within their structures. For them, the supreme court of decision is revelation, not logical reasoning. For instance, consider the fact that it is possible to postulate a logically impossible trinity, but not the existence and nonexistence of a God at the same time—although in both cases logic is similarly suspended. The "strictly logical" heresy in the Judas story means that his postulated "role as savior" is proved by the same logical means that belong to the arsenal of the traditional demonstrators of Christian theology. The heterodoxy arises only because Borges does not halt where, according to the Scriptures, any orthodox theological attempt at interpretation must "desist unconditionally." Borges's conclusions lead to a point which transcends the permissible boundaries, but this does not destroy logic, for this boundary is of an extralogical nature.

stance, might really be possible. Such a perfidious interpretation of the myth of the redemption, if historically not
very plausible, is at least thinkable. I could say the same
about "Lottery." Under certain conditions even the rein-
terpretation of the notions of chaos and order shown here
may be historically plausible. Both stories, different as
they may appear to be from one another, are hypotheses
about the structure and attributes of existence. Because
they are both borderline cases, isolated to one edge of the
real paradigm corresponding to them, it was very unlikely
that they would come true historically. Yet, considered
from a logical point of view, they are totally "correct."
The author therefore has the courage to deal with the most
valuable goals of mankind just as mankind himself does.
The only difference is that Borges continues these com-
binatory operations to their utmost logical conclusions.

Borges's best stories are constructed as tightly as mathematical proofs. It is impossible to refute them logically,
however lunatic the stories' premises may sound. Borges
is successful because in any single case he never questions
the implied premises of the model structure that he transforms. For instance, he pretends to believe (as some humanists do) that a truly brilliant work of art contains no
trace of chance, but is indeed the result of some (higher)
necessity. If one thinks that such a statement is generally
true, it is possible, without contradicting logic, to claim
that a masterpiece could be created, word for word, a
second time, and quite independently from its first birth
(as one can really do with mathematical proofs). We can
only see the nonsense of such a procedure when we attack
its very premises; but of course Borges is careful never to

do this. He never creates a new, freely invented paradigm structure. He confines himself strictly to the initial axioms supplied by the cultural history of mankind. He is a mocking heretic of culture because he never transgresses its syntax. He only extends those structural operations that are, from a logical point of view, "in order," i.e., they have never been seriously "tried out" because of historical extralogical reasons—but this is of course another matter altogether.

Basically, Borges just does what he claims for the fictitious philosophers of his "Tlön" (in philosophy they "do not seek truth, only amazement"). He cultivates a fantastic philosophy, for the characters and settings in his stories are not discursive arguments, but just as much literary objects as the objects which appear in "normal" literature. This group of tales forces me to ask how we may distinguish a fictitious ontology (one that cannot be taken seriously) from a real (historically valid) philosophy. The answer to this question is shocking: no essential difference separates the two. Things are quite trivial: those ontological-philosophical concepts that some thinkers had, and that were preserved by mankind in her historical treasure trove of ideas, and that she therefore acknowledges as serious attempts to interpret and understand the world in one grand sweep—those ideas are our religions and philosophical systems.[3] But ideas that cannot present such a

---

3. If Schopenhauer had never existed, and if Borges presented to us the ontological doctrine of "The World as Will," we would never accept it as a philosophical system that must be taken seriously; we would take it as an example of a "fantastic philosophy." As soon as nobody assents to it, a philosophy becomes automatically fantastic literature.

genealogical attestation and cannot show such an assimilation by the real history of mankind (and Borges's cannot) are just "fictitious," "freewheeling," "privately invented" meaningful structures, and for no other reason than that mentioned above. Because of this, they can never be taken seriously as an interpretation of the world and existence. These stories cannot be refuted even when the most severe criteria are applied, but only because things happen to be so. To refute them, it would not be sufficient merely to show their absurd consequences. To refute them, it would be necessary to call into question the total syntax of human thought, and thinking in its ontological dimensions. Therefore, Borges's work just confirms that no cultural necessity exists in our growth toward knowledge; for we often take that which has arisen by accident for what is necessary, and mistake the ephemeral for the eternal.

I'm not sure whether Borges would agree with my explication of his work, but I do fear that I have attributed more to him than he deserves, and that he has not written his best work with so serious an intent (in its semantic depths, not its comical-paradoxical surface, of course). Which means that I suspect that Borges "privately" has not seen the final point of his fictional chain of proof. This guess is based on a knowledge of *all* of his stories. By talking about his other stories, I pass onto the other, more dubious aspects of his work. Seen as a whole, his work is a universe of literature whose secondary, repetitious aspects diminish and slight his best efforts by their very neighborhood, because these aspects structurally debunk his best work. In Borges's best stories one can find flashes of such an intellectual power that they do not lose impact even after many rereadings. If at all, they are lessened only when one reads all of his stories at a sitting.

Only then do we notice the mechanism of their creative process. It is always dangerous, even fatal for the creator, when we see the invariant (debunking) structure, the algorithm of his creative power. God is a total mystery to us above all because it is on principle impossible for us—and will remain impossible for us—to understand or imitate exactly the structure of God's act of creation.

Considered from a formal point of view, the creative method of Borges is very simple. It might be called *unitas oppositorum,* the unity of mutually exclusive opposites. What allegedly must be kept separate for all time (that which is considered irreconcilable) is joined before our very eyes, and without distorting logic. The structural content of nearly all of Borges's stories is built up by this elegant and precise unity. Borges calls the one and the same the conflicting notions of the orthodox and the heretic, Judas and Jesus. Christ, betrayer and betrayed, the troglodytes and the immortals, chaos and order, the individual and the cosmos, the nobleman and the monster, good and evil, the unique and the repeated, etc. His literary game with its borderline meanings always begins where opposites repel one another with their inherent force; and it ends as soon as they are joined together. But we can see a trivial weakness in Borges's work in the fact that there is always the same mechanism of conversion (or a closely related inversion). God the Almighty was wise enough never to repeat Himself in such a manner. We authors, his successors, shadows and apprentices, also mustn't do it. Occasionally—but very

rarely—the skeletal, paradigmatic structure of the transformations used in Borges's fiction results in truly extraordinary things, as I have tried to show. But we always find
this structure, invariably in the same form, once we have
properly recognized and assessed it. Such repetition, which
inevitably is already accompanied by an element of the
unintentionally comical, is the most familiar and most
general weakness in all of Borges's fiction. For as good old
Le Bon has already said in his work on humor, we always
look disdainfully down upon the mechanic, for a mechanical process always lets the strange and surprising get away.
It is simple to predict the future of a purely mechanical
phenomenon. In its utmost depths, the structural topology
of Borges's work acknowledges its relationship with all
mechanistic-determinist kinds of literature, including the
mystery novel. The mystery novel always incorporates
unequivocally the formula of Laplacean determinism.

The cause of his work's "mechanistic" sickness is this,
I think: from the beginning of his literary career, Borges
has suffered from a lack of a free and rich imagination.[4]
In the beginning he was a librarian, and he has remained

4. This can be seen from the fact that several times he has rewritten
material supplied by others. I have not discussed this aspect of his
work, for I believe that nothing can be more erroneous in criticism than
to descend to the shallow passages of the work of a writer merely in
order to prove their worthlessness. Besides, it is an undisputed fact that
world literature is full of prose that is similar, and the immense number
of such exercises alone deprives of originality any piece that can defend
its individuality only by stylistic means. You can see this in the stories
comprising the last two parts of the Hanser volume [which Lem was
reviewing in this essay—TRANS.], especially in regard to the stylistic
means employed, whose baroque character is stressed by Borges in his

one, although the most brilliant manifestation of one. He had to search in libraries for sources of inspiration, and he restricted himself wholly to cultural-mythical sources. They were deep, multifarious, rich sources—for they contain the total reservoir of the mythical thought of mankind.

But in our age they are on the decline, dying off as far as their power to interpret and explain a world undergoing further changes is concerned. In his paradigmatic structures, and even in his greatest achievements, Borges is located near the end of a descending curve which had its culmination centuries ago. Therefore he is forced to *play* with the sacral, the awe-inspiring, the sublime and the mysterious from our grandfathers. Only in rare cases does

---

introduction. The more nearly a work becomes "literature," the greater its originality (as measured by the integral of its differences from all other literary works), the more this kind of fiction, which only increases the number of already existing texts by further similar elements, must be likened to the enlargement of an ocean by the pouring of water into it—it is, rather, a work of reproduction, more related to the crafts than to creative art. Of course, ninety-five percent of all writers are just craftsmen; but the historical movement of literature, and its historical changes, are caused by the inventors, heretics, visionaries, anticonformists—the revolutionaries of writing. And this gives us the right to measure any work claiming to belong to the top in literature first of all by its originality. Many writers can entertain; but only a few can amaze, educate, and move. But because such a point of view is open to attack, I have armed this review with a warning against its subjective character. Also I do not intend to evaluate the whole work of Borges, and especially not his poetry, which I would feel I would have to read in its original Spanish form. Whatever the matter may be with his poetry (which I value highly), it does not belong to fantastic literature for the simple reason that, in my opinion (and here I am in agreement with T. Todorov), on principle there can be no fantastic poetry.

he succeed in continuing this game in a serious way. Only then does he break through the paradigmatically and culturally caused incarceration which is its limitation, and which is quite contrary to the freedom of artistic creation he strives for. He is one of the great men, but at the same time he is an epigone. Perhaps for the last time. He has lit up—given a paradoxical resurrection to—the treasures transmitted to us from the past. But he will not succeed in keeping them alive for any long period of time. Not because he has a second-rate mind, but because, I believe, such a resurrection of transitory things is in our time quite impossible. His work, admirable though it may be, is located in its entirety at an opposite pole from the direction of our fate. Even this great master of the logically immaculate paradox cannot "alloy" our world's fate with his own work. He has explicated to us paradises and hells that remain forever closed to man. For we are building newer, richer, and more terrible paradises and hells; but in his books Borges knows nothing about them.

Translated from the German
by Franz Rottensteiner

ABOUT THE STRUGATSKYS'
ROADSIDE PICNIC

There are subjects that cannot be entirely exhausted. For theologians, such a subject is God. How can one definitively report on something that is, by definition, inexhaustible; how, when the description presumes a limit, can one describe a Being which, in principle, consists of infinite qualities? In this case, various strategies have been used: a multiplication of general concepts—which, however, generates no precise picture; comparisons—but they necessarily reduce the divine attributes to the level of all-too-concrete categories; or a spiraling approach to the subject whereby a definitive determination is replaced by an approximation—which for that reason is likewise inadequate.[1]

The optimal strategy for theology has proved to be that of maintaining the mysteriousness of God. Yet rigorously to preserve that mystery, one would actually have to remain silent; and a silent theology ceases to be theology. The strategy therefore turned (in later—e.g., Christian—versions) into one operating on obvious contradictions. God the omniscient knew that from man as He had made him would come the Fall. Yet God created him free. If God was aware in advance that man would inevitably fall, then man was not free—which nevertheless is exactly what the theologian asserted he was. In this way dogmatically imposed contradictions create the very mystery before which reason must become silent.

An inexhaustible topic of fantastic literature is the reasonable, yet not human, being. How can a human author describe a being which is definitely gifted with reason, but which, with equally categorical certainty, is not human? The bare assertion of its reasonableness will not do, since

---

1. The foregoing essay, subscribed "July 1975" in the Polish original, has been translated from the text that appeared as the Afterword to Arkady and Boris Strugatsky's *Picknick am Wegesrand* (Frankfurt: Suhrkamp, 1981), pp. 189–215. The essay first came out as an afterword to the Polish translation of the Strugatskys' *Piknik na obočine: Piknik na skraju drogi* (Kraków: Wydawnictwo Literackie, 1977), pp. 265–88. For this last bit of information, and for his generous assistance in checking our rendering of Lem, we are indebted to Dr. Franz Rottensteiner, who, however, is not to be held accountable for our errors. We are equally grateful to Elizabeth Kwasniewski for her help and patience in correcting the English translation against the original Polish. —Robert M. Philmus.

the genre must work with facts. Here, too, fantasts have resorted to various strategies. The one that proved the best in theology—namely, preserving the mystery—cannot be applied in exactly the same way: aliens, after all, are not deities but material beings like us. The author who describes them with the aid of various readily apparent contradictions is thus requiring the reader to believe in something absurd; whereas it is not, after all, in the writer's power to establish no-matter-what dogmas.

According to the simplest available strategy, then, intelligent beings differ from each other corporeally, and only from this area do their peculiarities arise. Mentally they are identical or similar to human beings, since there can be only one form of Reason. H. G. Wells gave reality to this view almost a hundred years ago in *The War of the Worlds.* His Martians have a horrifying appearance, which, however, will some day be man's. Their bodies have deteriorated to such an extent that their heads are almost all that remains; and, according to Wells's surmise, in the man of the future as well, the organism's viscera will atrophy and the cranium expand. The novel says nothing about Martian culture, as if that, too, had wasted away and consisted of nothing but technical mastery and the equation of might with the cosmic justification of the state. In Wells the future thus simplifies both physiology and culture. His Martians have no interest in anything human except human blood: like vampires, they nourish themselves on it. The Martians' technological achievements do, to be sure, arouse our admiration, but the poverty of their culture represents the fiction's greatest weakness. Let us not speak of the loathing the aliens inspire—that can always be referred back to their physical environment. Still, is the behavior of the Martians not an unintentional caricature of an extreme rationalism?

The invasion of Wells's Martians is certainly justified by their situation as inhabitants of a dying planet that is turning into a desert, from which perspective the fruitful earth hovers as territory *(Lebensraum)* to be conquered. What proves to be an exceptional case within the solar system was nonetheless thoughtlessly appropriated as the model for the whole science-fiction genre. Indeed, the successors to Wells mechanically imitated the failings of the master. The science fiction that followed his sickened on the chronic monstrosity of stellar invaders, while leaving behind the rationale by which Wells accounts for it. Furthermore, later writers, wanting at all costs to surpass the founder of the genre in their rendering of aliens' hideousness, went well beyond the limits of plausibility. By equipping their aliens with ever greater power, they filled the entire universe with civilizations whose desire to expand is wholly irrational. The greater the power attributed to the aliens, the more irrational is their invasion of earth. In this phase, science fiction became a fantasy of imposture and of paranoid delusions, because it claimed that the cosmic powers were sharpening their fangs the better to eat humanity, as if earth and its treasures were of incalculable value not only for the inhabitants of a small desert planet like Mars, but also for every imaginable civilization in the galaxy. Yet the preconception that a power with armies of starships at its disposal could be dead set on taking over our property is as naïve as the assumption that one of the superpowers of earth would mobilize its armies in order to expropriate a grocery store. The price of the invasion must always be higher than the value of the loot.

Thus invasion plots could not be motivated by interest in material gain. Instead, the aliens attack earth because it pleases them to do so; they destroy because they want to destroy; they enslave humanity because it amuses them to exercise tyrannical mastery. In this way, science fiction exchanged Wellsian interplanetary Darwinism for a sadism which became a cosmic constant in intercivilizational contacts. Science fiction's task of forming hypotheses was replaced by that of projection, in the sense the word has in depth psychology: the authors projected their fears and self-generated delusions onto the universe. They thereby established a paranoid cosmos, in which everything having so much as a hint of life sets about the conquest of earth—a cosmos that is a trap set to catch humankind, a cosmos whose evolution comes down to an embodiment of the principle of "Civilization as a wolf to Civilization" (cp. *homo lupus homini*).

This "den of thieves" cosmos was later transfigured many times over. Its general unfriendliness was mechanically transformed into friendliness. The aliens attack, but only to rob us of our free will and to preserve humankind by taking us into protective custody (this motif became especially popular during the Cold War years); or they don't attack immediately, but hesitate and thus enable humankind to unite: in view of the stellar threat, solidarity wins.

Further permutations of the invasion scenario resulted from these; yet none of the variations invented stands up to a thoughtful examination. They are incapable of answering certain elementary questions that Wells's novel—albeit in its own way—does pertinently address itself to. There is, for one, the question of what the motive is for the star-voyage—something that cannot be explained in terms of "they felt like it" or of a game of cops and robbers; for another, there is the question of the main orientation of cultures on a high level of material development; for yet another, there is the question of what form systems that have achieved a high level of astrotechnical accomplishment will assume; and so forth. But the most telling of such questions is this: why do actual human cultures show a tremendous richness approaching the truly diverse, while virtually all cosmic cultures in science fiction are marked by a depressing uniformity which borders on monotony?

To such questions science fiction could make no answer as long as it exchanged reflection on the fate of reason in the cosmos for sensational stereotypes of interplanetary adventure. In this way, science fiction's line of development—and this concerns the subject under discussion—became antithetical to that of science. At a time when scientists, beginning to discuss seriously the problem of how one might communicate with other civilizations in the universe, were formulating the hypotheses that Reason takes various forms and that not all possible manifestations of intellect need assume the human form, fantasy was already at the opposite pole from such thinking, driving the last remnants of realistic concepts out of its sphere through its undisguised borrowing from fairy tales. In its desire to furnish the aliens with ever greater power, it already ascribed protean abilities to them: such a being can, just by wishing it, transform itself into a tree, into part of a rocket, even into a human being. It can also take over a human body and control the human mind, thus in effect giving new life to a subject of old myths: possession by evil spirits. This fantasy destroyed intercultural barriers in short order, by ascribing some sort of telepathic omnipotence to the aliens; or, on the other hand, it formed the cosmic relationships between the planets on primitive, simplistic models of earthly origin (those, for example, suggested by colonialism, by the exploits of the conquistadors, or by the rules governing the creation of imperialistic coalitions). In so doing, it disregarded all possible objections both of a sociological and of a physical nature—objections that are contingent on the tremendous spatiotemporal distances in the cosmos. That handicap it did away with once and for all, by conferring on the starvoyagers the ability to move at any desired speed. In short, while in Wells's modest effort the Martians—in accordance with the scientific data of his time—were at home in the real cosmos, science fiction now chose to locate its beings in a totally (i.e., astronomically, physically, sociologically, and—finally—psychologically) falsified cosmos. It practiced a ruthless exploitation, ransacking, in its search for inspiration, history textbooks and the Linnean system alike, in order to provide lizards, cuttlefish with grasping arms, crabs, insects, and so forth with intelligence. When even that had become threadbare and presently boring, the theme science fiction had run into the ground was in its teratological extremism taken over by the third-rate horror movie, which is perfectly bare of any thoughtful content.

American writers deny the validity of such a diagnosis of the facts, and they find allies in the book buyers, who have become used to an easily digestible, sensationalistic literature that pretends to be science fantasy. Yet the fairy-tale nature of this "fantasy" is obvious. Nobody questions why the dragons in fairy tales are so mischievously blood-thirsty or why the witches in them prefer to devour children rather than chickens. These are simple axioms of the fairy tales, whose world is fundamentally partisan: evil appears in it so that it can be defeated by good. It is therefore clear that such evil must be powerful; otherwise, the final victory of good would seem too easily gained. The world of science fiction, on the contrary, must be impartial; it must not incubate evil merely for the sake of allowing the united interplanetary forces of virtue to overcome it. Nor should it be a partisan world with a minus sign, an anti-fairy-tale world in which the beautiful, amiable, and morally upright good is bred in order to give the greatest possible pleasure to an evil incarnate which proceeds to gobble it up with relish. (Such a world, incidentally, was imagined by the Marquis de Sade, whom one could hardly take for an author of science fantasy.) The science-fiction world must be (to put it quite plainly) a real world: that is, one in which no one is privileged from the start, in which no fate is predetermined, whether in favor of good or of evil. Since men are not angels, there is no need to ascribe angelic traits to the aliens; since men, though they kill flies, do not exactly travel to the ends of the earth to do so, similarly the aliens, even if they should regard us as flies, should not go out of their way to seek earthlings to swat.

An author who describes a life form or type of intelligence different from the terrestrial variety is in an easier position than the one who depicts a cosmic invasion of earth. The former can—as, for example, I did in *Solaris*—restrict himself or herself to portraying phenomena that differ as much as desired from what humans are familiar with. The latter, proceeding from the "interventionist" premise, assumes that the aliens have come to earth and that, consequently, something or other must have dictated their literally astronomical undertaking. What could their motive have been? If it was not an impulse to fight or to steal, it must have been the urge either to learn or to play (they came in order to amuse themselves a bit with us . . .). There are, as we see, not many alternate possibilities. Thus the best strategy for dealing with this subject, too, is to preserve forever the aliens' mysteriousness.

I would like to stress emphatically that this strategy is not founded, either entirely or primarily, on aesthetic criteria; that, in other words, the narrative must not preserve the aliens' mysteriousness in order continuously to puzzle readers and hold them spellbound by the great unknown. The strategy does, of course, incline to conform to the fundamental directives of conflict theory. Thus, by way of example, future-strategists at military academies are required to impute to the enemy the most threatening intentions from the point of view of the strategists' own side. In regard to cosmic aliens, such a dictate has a cognitive, rather than a military, purport. Yet visitors fitted with absolutely inimical intentions do not represent the worst of all possible eventualities. In this case, the enemy's attitude is at least clearly defined. The situation is worse when we absolutely cannot understand the peculiarities of their strange behavior, when we cannot explain their alien proceedings.

The strategy of preserving the mystery, if it is to be optimal, requires a precise concretizing. One cannot manage it in the way that theology does its subject, by working with contradictions. One cannot ascribe mutually exclusive purposes to the visitors—for example, they cannot want to conquer and at the same time not conquer. Still, one can rouse the appearance of such a contradiction—for example, the visitors may believe they are helping us, though we may feel that their actions are pernicious—and here one enters the realm of what is promising from a dramaturgical perspective: misunderstandings occasioned by the drastic disparity between civilizations. One can find attempts in this direction in science fiction, but they are not followed through: the intercivilizational misunderstandings always stay extraordinarily primitive puerilities which do not merit serious consideration. The author must invest a certain amount of intellectual effort in the construction of the quid pro quo that perplexes the meeting of two disparate cultures. The more factors from various areas that contribute to such a misunderstanding, the better. One ought to keep in mind that such an encounter is not a duel between two heroes, but a very confused interplay in which collective social organizations take part, organizations that differ radically from each other and to each of which the structure, meaning, and purpose of the other's actions are foreign.

The overwhelming majority of science-fiction texts can serve as examples of how not to tackle the theme of invasion. It is therefore all the more gratifying to come upon a work which, by and large, knows how to deal with the problem successfully. In *Roadside Picnic,* the Strugatsky brothers have employed the tactic of preserving the mystery to excellent effect; indeed, as they surpass the canon established by Wells, so, too, they transcend the science-fiction tradition.

Roadside Picnic relies on two ideas. The first we have already designated as the strategy of preserving the mystery of the visitors. One does not know what they look like; one does not know what they want; one does not know why they came to this world, what their intentions were respecting humankind. Nor does one know exactly whether it's absolutely certain that they have landed on earth at all, and if they have, whether they have already left again. . . .

The second idea—and this is what makes *Roadside Picnic* an anomaly in science fiction—pertains to humanity's reaction to the landing. For something has landed—or, to put it more circumspectly, something has fallen from the sky. The inhabitants of Harmont have found that out tragically enough. In some areas of the city people go blind; in others, they fall victim to mysterious illnesses that are generally described as plague; and the depopulated area of the city turns into the Zone, whose properties, menacing as they are incomprehensible, abruptly separate it from the outside world. Yet the actual landing was no great natural catastrophe: it did not cause houses to topple down, nor did it make windows break for miles around. The book does not tell us much about what happened in the first phase of the creation of the Zone. Still, we learn enough to understand that we will not be able to fit the events and their consequences into any compartment of already-existing classificatory schemes. Those who escaped from Harmont in one piece and moved elsewhere become the center of incomprehensible events, of extreme deviations from the statistical norm (ninety percent of the clients of a hairdresser who left Harmont die in the course of a year, though of "ordinary" causes—in a criminal attack, in traffic accidents—and wherever emigrants from the Zone increasingly congregate, the incidence of natural catastrophes rises proportionately, as Dr. Pilman informs Noonan).

We thus have before us an incomprehensible infringement on causal connections. The narrative effect is striking. It has nothing to do with phantasmagoria in the form of a "visitation," because nothing supernatural occurs; and yet we are confronted with a mystery that is "much more terrifying than a stampede of ghosts" (as Dr. Pilman says, 3:109).[2] Should someone seek for a hypothesis that would explain these effects, it might be possible to find one. (Let us assume that what has happened is caused by local disturbances of certain physical constants responsible for the normal probability curves in typical statistical equations: that is the easiest explanation, though only, of course, as it indicates the direction in which more research would have to be done, and not in the sense of being a solution to the problem.) It turns out, then, that even

---

2. Parenthetical references (which the translators have supplied) are to *Roadside Picnic*, trans. Antonina W. Bouis (New York: Macmillan, 1977). This volume also includes *Tale of the Troika*. —Robert M. Philmus.

when one has found a physical process whereby the mechanics of the unusual events can be explained rationally, one has not come a hair's breadth closer to the heart of the problem—viz., to the nature of the visitors. Thus the optimal strategy consists of presenting the individual actions of the visitors as a puzzle whose resolution either does not throw any light at all on the nature of the visitors or makes that nature seem even more unfathomable. This is not, as it might perhaps appear to be, something made up, like a fantasy novel's ad hoc inventions; since our knowledge of the world is acquired in just this way: perceiving some of its laws and peculiarities does not lessen the number of problems left to be solved; on the contrary, while making these discoveries, we begin to realize that there are further mysteries and dilemmas of whose existence we hitherto had no presentiment. Evidently, then, the scientific learning process can produce from its treasury even more "fantastic" wonders than the fairy-tale repertory does childish ones.

In *Roadside Picnic* things do not go as they do in *The War of the Worlds.* Wells's story of the Martian invasion involves a nightmarish, monumental breakdown of the human world, a dramatically heightened collapse of civilized order under visibly inflicted blows. One knows who the opponent is; one knows his methods; even his final goals are known (it would be difficult not to guess them!). All this has nothing in common with *Roadside Picnic.* To be sure, the invasion has presumably occurred; to be sure, it has left behind ineradicable traces in the form of "Zones"; and earth is incapable of coming to grips with the consequences. Yet at the same time, the little world of humanity continues as before. Ominous miracles, descending on six spots on the planet like a cosmic rain, become the focal points of the various—legal as well as illegal—human activities that go on around all supposed sources of profit, no matter how risky they are. The Strugatskys realize the strategy of preserving the mystery through an extremely subversive tactic—through well-nigh microscopic bearings on what is going on. We learn only through hearsay that experiments are being made on the "magnetic traps" discovered in the Zones, and that somewhere or other institutes for the study of extraterrestrial cultures are busy trying to comprehend the nature of the landing. About what governments think of the Zones, about how the Zones' instauration has affected world politics, we find out nothing. By contrast, we witness every last detail of some episodes in the life of a "stalker," of a new breed of smuggler who, because a demand exists for them, spends nights retrieving objects from the Zone. Through verbal snapshots, the story shows how the Zone has become surrounded, as a foreign body does when it has penetrated a living human organism, only in this case by a tissue of opposed interest groups: those connected with the official guardianship of the Zone (i.e., the UN), but also the police, the smugglers, the scientists, and—let's not forget them—the members of the entertainment industry. This encirclement of the Zone by a ring of feverish activity is depicted with considerable sociological insight. Certainly the portrayal is one-sided, but the authors had good reason to focus on those figures whose activity, in a marked but also quite natural way, counters the typical science-fiction scheme of things. The sense of fascination and depression that the "scenes from the life of a treasure hunter" (or "stalker"), the core of the story, inspire in the reader are the product of a deliberately restricted field of vision. The scientific and extrascientific literature that the landing precipitated must undoubtedly have been a locus of bitter controversy. So, too, the landing must inevitably have brought about the formation of new attitudes and lines of thought; and it probably has not left either art or religion untouched; yet our perspective on the whole upheaval is perforce confined to the excerpts from the life of a poor joe who, in the drama of two civilizations colliding, strictly plays the part of a human ant.

It would nevertheless be a good idea for us to make ourselves aware of wider aspects of the event. Everyone will agree with Dr. Pilman's words that the invasion represents a decisive stroke in the history of mankind. Now in that history there have been quite a number of decisive moments, even though they were not exactly caused by a cosmic invasion; and each was marked by an intensification of the extremes of human behavior. Each of these decisive moments had its larger-than-life-size figures and its pitiable victims. The greater the historic event, the more pronounced was the distance between the great and the insignificant, the sublimity and the wretchedness of human fates. Glorious battles at sea that once decided the destiny of empires possessed at a distance the beauty of a painting of a battle, and close up a repulsive gruesomeness. One need only recall that chained to their benches, the rowers of galleys burned to death in Greek fire silently, because before the battle they were obliged to stuff their mouths with special pears to prevent their making any noise. (Their hellish shrieks, you see, would have had a negative effect on the soldiers' morale!) Perceptions of such a battle would differ radically, depending on whether it were seen from the elevated perspective of the commanders, with their imperial aims, or from the viewpoint of the poor devils faced with a death struggle—and yet their death struggle was an integral part of the process of historical change. One could say that even such a beneficial discovery as that of X-rays, for example, had its horrific side, since the discoverers, unaware of the properties of these rays, had to have limbs amputated because of their effect. So, too, one of the by-products of the world's industrialization is the leukemia that children are slowly dying from today. (We know this to be true, even though the causal connection cannot be palpably demonstrated.) The dreadful fate of the stalkers in *Roadside Picnic,* I should add, does not represent an extraordinary deviation brought about by the cosmic landing, but is precisely the rule of decisive moments in history—a rule that distinctly points up the constant and inevitable connection between "picturesque" greatness and horrifying misery.

The Strugatsky brothers thus demonstrate that they are realists of the fantastic inasmuch as realism in fantasy betokens a respect for logical consequence, an honesty in deducing all conclusions entirely from the assumed premises. Even the uninhibited entertainment industry that encircles the Zone has its plausibility—indeed, I would say its necessity. The principles of human behavior operating in the narrative are thus the same as they ever are; the authors have merely directed their attention to the "dregs" in the cosmic encounters, so to speak—and the

events thereby take the concrete form of a miracle introduced into a consumer society. This is not—what is sometimes meant by the latter term—a society that produces
nothing but those goods to which consumers are immediately attracted. On the contrary, it is a society that considers everything to be within the scope of its endeavor: not
only cars, refrigerators, and perfume but also sex, blood,
and destruction it makes items for consumption, in good
time seasoning each of them so that they become palata-
ble. In the Middle Ages, the Zones would doubtless have
caused movements of panic-stricken flight and migration;
and they might afterward have become centers of new
religious beliefs, originating in response to their evidence
for the Apocalypse, and breeding grounds for prophecies
and revelations. In our world, however, the Zones suc-
cumb to being domesticated; for what one can neither
understand nor ignore, one can at least consume piece-
meal. Accordingly, the Zones, rather than being the subject of eschatological thought, are the goal of bus tours.
This admits of being explained with reference to a lust for
phenomena once regarded solely as abhorrent but these
days enforcing the popularity of an aesthetic that in place
of beauty has set the repulsive. That is the spirit of the
times to which in the Strugatskys' story anything evincing
its complete independence from man—as the mysteriousness of the visitors nearly does—succumbs. All in all,
Roadside Picnic implies that the landing passes over ninety-nine percent of humankind without a trace; and precisely in this regard, the Strugatskys set themselves against
the entire science-fiction tradition.

Theirs is no banal opposition. Dr. Pilman, given to

expressing himself in the terminology of physics, calls mankind a "stationary system" (3:100); translated into the language of the historian, this means that contact with the aliens, insofar as it does not equate with a global catastrophe, cannot change the course of human history with one fell swoop, since mankind is not capable of suddenly leaping out of its history and—impelled by a cosmic intervention—stepping into a completely different history. This supposition—in my view a correct one—is something that science fiction has neglected in its avidity for the sensational. In *Roadside Picnic,* by contrast, the landing is not intended as something strange for the sake of its strangeness; instead, it establishes the starting conditions for a thought-experiment in the domain of the "experimental philosophy of history"—and that is exactly what determines the value of this book.

There is only one point about which I would fain take issue with the book—a point having to do not with human matters (these are presented unobjectionably), but with the actual nature of the visitors. I might premise my discussion on four propositions. The first is that in the book we are given data, but not necessarily opinions about these data, even when the characters harboring such opinions are holders of the Nobel Prize. This means that we consider ourselves to have as much of a warrant to postulate theories about the visitors as the fictional personages have. Second, on all imaginable levels of knowledge there is given no hundred-percent error-free course of action. Such infallibility would of course require that one possessed complete information about what can occur in the course of making one's plans a reality; but the universe is a place in which the attaining of complete information about anything whatever is never possible. According to the third of my propositions, the principle of freedom from contradictions in thinking obtains for us as for other beings in the cosmos. This means that of two things, one must prove to be the case: if the "visitors" were aware of the presence of humans on earth, then they cannot at the same time have been unaware of it; if they harbored any design at all vis-à-vis human beings, then they could not at the same time have harbored no design; and so on. Finally, in explaining unknown phenomena, the simplest hypotheses, as stipulated by the principle of Occam's razor, are always to be preferred. If, for example, we live next door to a famous magician and hear dead silence from his side of the dividing wall for a long while, we can explain that in many ways: the neighbor may have dissolved into thin air, or he may have transformed himself into a paper clip, or he may have gone up to heaven through his window. We would tend to take refuge, however, in the quite commonplace explanation: that he simply left the house quietly and unperceived. Only when that hypothesis proves wrong are we compelled to look for another, and less banal, one.

These are the standpoints from which we negotiate the encounter with the visitors.[3] In regard to the landing, a

---

3. Our entire explanation, which provides a new interpretation of the riddle presented by the landing in *Roadside Picnic,* may seem to be an aberration brought on by excessive pedantry, all the more so since, after all, we are not analyzing a real event, but a literary fiction. But truly scientific fantasy is distinguished by just this: that one can subject the events described in it and their rational depiction to the same proof of

distinction must be made between what the aliens left in the Zones and the way in which they did this. In the opinion of Dr. Pilman, who expresses the outlook of most specialists, the gap between the civilizations turned out to be too great for human beings alone to be able to surmount it; the other side, however, failed to give its assistance. What the visitors have left behind human beings can deal with only as fragments of a strange technology whose functioning is incomprehensible. As for the manner in which the visitors bequeath the so-called objects to men, Dr. Pilman's thesis—which is central to the story since the title on the cover already anticipates it—represents this to us in the form of a parable. Mankind finds itself in the

---

coherence as phenomena that occur in the extraliterary world. Such a work may start with a fictive, even an extremely fictive, premise. Yet this authorizes only an initial poetic license, which loses its validity within the story itself. This means that the storyteller may not, within the story, continue to help himself along by the ad hoc invention of whatever things or phenomena strike his fancy. Fairy tales may operate with such ad hoc inventions; for they are not at all required to explain logically or empirically the miraculous occurrences they depict. A science-fiction story that makes this fairy-tale license its own leaves the realm of the real world and puts itself in the position of the fairy tale, in which everything that is thought of or said for that very reason instantly becomes possible and must be unquestioningly accepted as true coin by the reader. In short, though the facts in a science-fiction story may be fictive, the way in which science in the fiction interprets these facts may not. Scientific theories change; but what does not change is the method of discovery that characterizes science, and it is precisely this methodology that dictates a certain type of hypothesis-formation in science fiction. Accordingly, our polemic, as an example of how criticism of science fiction should typically proceed, can be applied *mutatis mutandis* to every work that fulfills the main criteria of this genre.

situation of animals which, having crept from their hiding places to a roadside or clearing where incomprehensible creatures have stayed, rummage around among the remains of the campsite. This analogy expresses Pilman's honest conviction, even though in his conversation with Noonan he enumerates other going hypotheses about the landing. Dr. Pilman is not just anyone; he has finally received the Nobel Prize for his discovery of the "Pilman Radiant." At the same time, he is a misanthrope—as outstanding scholars frequently are. Such men strongly sense the ambiguity of their societal role. For civilized society, which is brought forth from the fruits of their thinking, they are indispensable; yet it treats them quite inconsiderately. The political powers expropriate their discoveries, but public opinion nonetheless makes the researchers themselves answerable for the consequences of that expropriation. An awareness of this situation does not dispose one toward kindliness. Instead it arouses either rebellion or cynicism; but whoever finds rebellion useless and cynicism repugnant tries to behave like a stoic. Such a person gets used to choosing the lesser evil; and when one tries to corner him with questions, he answers evasively or with sarcasm. This is precisely Dr. Pilman's attitude, a primarily defensive stance which he has assumed in the interview with which the story begins.

In his conversation with Noonan, Pilman is certainly less spitefully laconic than he is with journalists. Because he is talking confidentially to someone he knows—besides which he is somewhat inebriated—he inclines toward straightforwardness. That Pilman, as he is psychologically delineated by his judgments on the landing, is by the same token not unbiased is another matter. The simile of leftovers from a picnic which he has availed himself of may accurately reflect the situation of human beings vis-à-vis the things found in the Zone; but in respect to the visitors it is all too lenient. The so-called leftovers, objects that are dangerous to all living beings, were not after all thrown away in some deserted spot. They were tossed into the middle of a city. It is a fact that urban areas do not amount to even one percent of earth's surface. That is why, though the cosmos has been "throwing" meteors at the earth for millennia, so far no meteor has fallen on a city. It would seem, then, that the landing in Harmont was not the work of chance. One could suppose that the visitors landed in the city because they wanted to. They held their picnic not on a roadside or in a deserted clearing, but right on top of our heads. The event thereby appears in another light. There is, after all, a difference between sitting down for a picnic near an ant hill, and pouring gasoline from the car over the ant hill and setting it on fire. The roadside picnic of Pilman's friendly analogy presupposes total indifference to the fate of the human ants. The picture of deliberate destruction, on the other hand, presupposes a high level of ill will, since one would really have to take the trouble of coming from far off in order to destroy the ant hill. Indifference and malevolence are not the same thing; and in this regard it is unfortunate that the story is silent as to whether even one of the other landings took place in a human settlement.

As we see, this bears on a paramount question, one that is critical for clarifying the visitors' relationship to man, and hence a question that all the characters in the fiction must similarly be aware of. One landing in a city could be the work of extraordinary chance, but two such landings certainly could not. We are thus inevitably led to the following reflection: if it were the case that the visitors had also landed in another city apart from Harmont, a roadside picnic would manifestly be a false image. But since Pilman has chosen this analogy, we assume that we are dealing with an isolated fact. That is very important for our further considerations.

Dr. Pilman details various hypotheses concerning the nature of the landing. He omits only one that commands our attention. This we will introduce after we have gathered (as follows) the conclusive evidence in its favor. . . .

(I) First of all, two characteristics—unrelated to one another—of almost all the objects found in the Zone attract our notice. One is that these objects have retained a measure of functionability: they are not passive, lifeless, deactivated waste or rubbish. The second is that they are commensurable in size (and weight) with the human body. This can be inferred from the fact that one man can lug objects well-nigh intact from the Zone on his back without having to take them apart first. None of the larger units needs to be dismantled or broken up—which is why the equipment of the treasure hunters includes no tools. These objects lie loosely strewn about. Now suppose we were to dump a considerable amount of the industrial debris of our civilization (wrecked cars, industrial equipment, scrap metal, old bridges, used machines) here and there on the Samoan Islands: the natives would in that case come upon far more objects incommensurable with their bulk than corresponding to it. If, on the contrary, a number of strange objects found scattered in a given place were of an order of magnitude according with that of the human body, it would be an a priori probable hypothesis that these things had been destined for their discoverers. Naturally one can still claim that pure chance is answerable for the fact that the objects found in the Zone are proportionate to human bulk. But it appears otherwise when many "pure coincidences" begin to come together in a meaningful pattern.

(2) Next, it is notable among the Zone's numerous characteristics that its boundaries are rigidly and sharply fixed. Neither flying objects such as the "hairy stuff" (1:19) nor other Zone phenomena (the "jelly" [2:56], thermal shocks, etc.) ever cross over the demarcation line between the Zone and its environs. One could once more claim that this "self-containment" of the Zone, which sets its own strict limits, is the result of a further "pure coincidence." However, it is a priori a more probable hypothesis that this is not the case, but that the Zone "holds itself in check" because it contains something that, according to the visitors' plan and intent, merits such enclosure.

(3) Then again, entire objects lie scattered chaotically in the Zone. It is probably this that has put Dr. Pilman in mind of a roadside picnic, where garbage is left behind. It thus really does appear as if these things were carelessly thrown away. But one can also defend the view that nobody threw them away, that they scattered themselves chaotically when the containers they were brought in burst.

(4) Furthermore, the objects in the Zone frequently

have the character of extremely dangerous pitfalls or booby-traps. Compared to dealing with them, defusing bombs or mines is sheer child's play. Again one cannot exclude the possibility that they were carelessly discarded by visitors indifferent to human welfare, or even the other possibility that the visitors treated people in the way that an assassin treats children when he passes out poisoned candies in a kindergarten. But another explanation is also permissible: that the objects do not function in the way they should because they were damaged during the landing.

(5) Finally, it is worth remarking that among the forces at work in the Zone are those that produce an effect of "resurrection from the grave." Under their influence, human corpses rise up and begin to move about. This is treated as a resurrection not of the dead, who are thereby returned to their normal state of living, but—to use the term from the story—of "moulages— . . . dummies" (3: 109), whose newly formed tissue is not identical with normal living tissue. To quote Pilman: "If you cut off some part . . . [from these living corpses], the part will live on. Separately. Without any physiological solutions to nourish it" (3:109). (Dr. Pilman claims that such a quasi-resurrection would violate the second law of thermodynamics; this is not a necessarily valid conclusion, but we do not want to quarrel with the learned man at this point, because doing so would take us too far off our track.) The "pseudo-resurrection" of the "zombies," or "moulages"—their reconstruction from a skeletal basis—is an effect that is very important for understanding the nature of the landing. Their "resurrection" seems to be more probable as a

consequence of purposive rather than undirected activities; and by the same token, it would certainly appear easier to resurrect real, concrete life forms (i.e., terrestrial ones, consisting of albumen) than "omnipossible" (and hypothetical) forms of life in the cosmos. We do not know if that is correct, just as we do not know if the effect was not directed exclusively at the visitors themselves (it might be a "remedy in their first-aid kit"). But whatever is the case, the resurrection effect suggests that the visitors knew a lot about the physiology of terrestrial life forms.

All of this constitutes the evidence for our hypothesis. We maintain that there has been no landing after all. Our hypothesis, indeed, runs otherwise. . . . A spaceship filled with containers that held samples of the products of a highly developed civilization came into the vicinity of the earth. It was not a manned ship, but an automatically piloted space probe. That is the simplest explanation of why no one manages to observe a single visitor. Every other hypothesis has to assume either that the visitors are invisible to humans or that they deliberately hide from them. In the approach to earth, the vessel sustained damage and broke into six parts, which one after another plunged from their orbit to earth.

This seems to contradict what is said about the radiants discovered by and named after Dr. Pilman; they ostensibly confirm that Someone fired at earth six times from Alpha Centauri in the Cygnus constellation. Nevertheless, between our interpretation and the radiants there stands no contradiction. In astronomy the term "radiant" refers to a likely place in the heavens from which a meteor swarm is approaching. The determination of a radiant in astronomy is not synonymous with taking a fix on the place from which the meteors actually originate. They may approach on an elliptical or parabolic curve; the radiant is a tangent which a terrestrial observer plots on such a curve, and it extends backward (in the opposite direction from the meteors' fall) until it reaches the place in the heavens where a specific cluster of stars is located. Thus when one names meteors according to their radiants, this by no means signifies that the meteors are in fact emanating from that star cluster after which astronomers have named them. Consequently, the Pilman Radiant offers us no clue at all as to whether what descended in the Zones was actually sent from the principal star of the Cygnus constellation. As to where the six flying objects or probes came from, the Pilman Radiant can tell us nothing, albeit the story directly encourages the impression that it can. That is a false impression, occasioned by Dr. Pilman's insufficiently precise manner of expressing himself in responding to journalists' queries at the outset of the story. That what has arrived here actually flew directly from Alpha Centauri to earth is out of the question. Traversing such a distance on a perfect flight path borders on an astronautical impossibility, since on the way innumerable interferences (above all, those of gravitational forces) must influence the trajectory. In addition, it is mathematically demonstrable that the curve that six shots would produce on the surface of a sphere (while the sphere is rotating, as the earth does) cannot be distinguished from the curve resulting from the projection of a segment in the orbital trajectory onto the surface of the sphere. Pilman's radiant does not altogether exclude the hypothesis of a disintegrated spaceship crashing in six separate sections, one after the other. Once one knows a meteor's radiant and its final velocity, one can compute the orbital path from which it actually approached, because a meteor, being an inanimate object subject to the laws of celestial mechanics, cannot alter its course at will. From a spaceship's radiant, on the other hand, one can make out nothing about its place of origin, its course, its travel speed, etc., because a spaceship is a navigable, mechanized body and can execute maneuvers, make course corrections, change its speed, and so forth. In short, from the so-called Pilman Radiant nothing follows in favor of any one of the hypotheses about the landing.

Of course we cannot know for sure that the spaceship was indeed the victim of a catastrophe. Nonetheless, our hypothesis accounts for everything that happened, and does so in the most economical manner. Why should one not properly assume that the landing has miscarried? To suppose that the unusual nature of the objects found in the Zone demonstrates the high level of the visitors' ingenuity and thus precludes a calamity's befalling their ship is a logically false inference. The visitors' perfection, in consequence of which no harm could come to their ship, is neither a fact nor a rationally defensible hypothesis, but merely an article of faith. Perfection to the point of infallibility is, in our judgment, reserved solely for those entities with which theology concerns itself—which is to say, there is no infallible applied science. We are not asserting that an accident definitely occurred, merely that a breakdown would, in one fell swoop, account for everything that happened by reference to a common single cause.

Besides, the facts that we mentioned in point (1) above

about the characteristics of the objects found in the Zone make plausible the conjecture that someone sent containers of technological specimens in earth's direction. Our point (2) (concerning the Zone's "self-containment") further increases the likelihood of (1), that the senders, unable to be absolutely certain that no catastrophe would befall their spaceship during its landing, must at least have provided for a minimizing of the consequences, and have done so precisely by installing on board a safety device that would not allow the effects of the catastrophe to spread, but would almost hermetically confine them to one place. This must naturally have been a device meant to survive the aftermath of the catastrophe. Somehow it has survived. The fact mentioned above under my third point heightens the probability that an accident has occurred, because nothing is more natural than that the containers' contents should be chaotically scattered by the force of the impact with earth. Even the fact cited as my fourth point (i.e., the perils the objects present) becomes understandable as a consequence of the same cause. Not only did the containers burst upon impact, but most of their contents were damaged in various ways. The same thing would happen if someone were to drop containers with foodstuffs, medicines, insecticides, etc., down to the Samoan Islands in what turn out to be defective parachutes. These crash to earth and the containers rupture—in consequence of which, the chocolates are full of hexachlorides, the gingerbread full of emetics, and so on. It is possible for the Samoans to conclude that someone has made a very malicious attempt on their lives; yet in the Samoans' place scientists ought not to jump to the same conclusion. What we are getting at is that the intention of the "Others" does not manifest itself in the pernicious character of the cosmic offerings: it is not the case that they took pleasure in pelting us with deadly debris, but, rather, that an unfortunate accident—the defect in their spaceship—transformed their well-intended consignment into scrap metal. (We do not want to go into further specifics of our hypothesis here; but in general terms, they would run as follows: since the ship left no trace behind, it must certainly not have landed but simply accomplished the dropping off of the containers. The containers, moreover, need not necessarily have taken the form of material vessels; the objects may have been "packaged," held together, by a type of force field, whose failure at a crucial moment caused the contents of the "packages" to rain down on earth.)

The Strugatskys might tell us that the hypothesis of "samples" is likewise taken into account in their book. After all, in his conversation with Noonan, Dr. Pilman makes mention of the possibility that "[a] highly rational culture threw containers with artifacts of its civilization onto Earth. They expect us to study the artifacts, make a giant technological leap, and send a signal in response to show we are ready for contact" (3:103). However, this version—which, by the way, does not admit the possibility that the consignments have arrived in a disastrously damaged condition—the story through an ironic undertone utterly discredits. Indeed, how could objects that are more dangerous than explosives and that are dispatched to unknown addressees as gifts be supposed to invite the recipients to make contact? That would be like sending someone an invitation to a ball, but enclosing the invitation in a letter-bomb. In the story's presentation of it, this hypothesis is therefore the one that is most self-compromised in view of the Zone's macabre characteristics.

The hypothesis of an accident, on the other hand, not only explains events quite naturally, but also rehabilitates at once the "Others" as Senders and the human beings as Recipients of a "Danaic gift" from the stars. The senders, far from being guilty of wrongdoing, have even—as was their duty—foreseen the worst possibility and provided the shipment with a safety device, thanks to which all of the Zone's effects are confined to a particular location. Accordingly, the Zone's peculiarity is most simply explained by the foresight of the senders, who, unable to eliminate the possibility of an accident, therefore took care that its consequences would be kept within bounds. The hypothesis of an accident likewise exonerates humankind, and especially the learned, whose perplexity about the gift becomes understandable, given the additional difficulties they have to overcome because they do not know which of the properties of the objects in the Zone were intended by the designers and which are the result of the damage incurred during the catastrophe.

It does not take long to explain why the authors silently passed our version of the landing by. It could not please them because it detracts from the work's menacing and hence mysterious atmosphere. Still, their error lies in just this silence about the possibility of an accident. We understand quite well why they chose this course. In the meeting of the civilizations, both sides were *meant* to be discredited. Men agree on using the gift only in base and self-destructive ways because that is human nature; and the Senders prove their murderous indifference to humanity because beings of high intelligence do not give a damn about their intellectual inferiors. So extreme a version of the invasion theme would have deserved literary representation, all the more since it surpasses everything that science fiction has so far accomplished in this direction. But in that case the narrative would have had to rule out our hypothesis about the damaged gift; it would have had to bring it to grief from the outset—that is, it ought to have discredited it. Silence about it, on the other hand, though intended to consign our version to oblivion, constituted a mistaken authorial tactic.

From what has been said, conclusions of a more general nature arise with regard to the optimal strategy for dealing with the invasion theme.[4] In order to carry out the strategy

---

4. We are presenting the hypothesis of the calamity in its simplest, which does not mean its most probable, version. For example, an unmanned spaceship with containers might have been sent forth without any fixed addressee in mind; it might have been outfitted with sensors that would recognize the planet to be "gifted" by virtue of predetermined parameters (such as its average temperature; its atmospheric composition, particularly the presence of free oxygen and water; an orbit favorable to ecological development; etc.). Such an automatically piloted vehicle could have approached various stars on a scouting mission. However, because it is physically impossible to manufacture technological products to survive undamaged over a journey of indeterminate length (which may take millions of earth-years), such a vehicle must have been provided with a device that would automatically destroy the contents when their "shelf life" had ended. Such a vehicle could have entered our solar system as the "shelf life" of the articles was nearing the expiry date. After all, it could also have been that the self-destruction did not occur only because the ship's surveillance system discovered earth and dispersed the containers with their "partially spoiled" contents. The degree of damage to individual surveillance, steering, and control systems is uncertain; only the statistical probability of damage can be determined—i.e., the one thing

of preserving the mystery, two requirements have to be rigorously fulfilled. First of all, the author must not arouse the suspicion that certain facts are being hidden from the reader, facts beknown to the fictive heroes (all of *Roadside Picnic*'s protagonists must know, for example, whether another Zone, aside from the one in Harmont, also lies within a city's limits). The reader must remain convinced that the information the author imparts is, within the limits of possibility, complete. The mystery then will be kept hidden by the very unfolding and presentation of the depicted events, which create, as it were, an impenetrable mask behind which no one can see. Otherwise this effect can be achieved only through a very precise balancing of the facts. They may neither point in one direction too unequivocally, nor overwhelm us by being all too chaotically diffuse. What they attest to must remain undecided, on the divide, as it were, between diverging alternatives, without inclining definitively toward any one side.

Now our excellent authors have defeated their own purposes by maligning the visitors at the end of their story. That the Golden Ball is supposed to fulfill wishes is, of course, a naïve belief, one of those popular legends that rose up in the wake of the visit. It was clear to the authors that they could not make an infernal machine out of this

---

absolutely certain is that the probability of defects occurring in the programs and their execution system increases with the passage of time. I should emphasize this point: the more complicated a device, the more inevitable are breakdowns over the course of time; this is a universal law that is independent of where in the cosmos the technology was produced or how. Therefore, the enterprise of learning about the aliens—what the Strugatskys call "xenology"—must take the statistical-probability aspect of intercivilizational contact into account as something crucial for interpreting such visits.

Ball, since that would have been an exaggeration that would have changed the meaning of their book: it would have transformed the Zone from something ambiguous, albeit dismal, into an unequivocal trap for humankind. Therefore they made the Golden Ball into an almost neutral object and let death stand not in it, but right beside it, as a "transparent emptiness that was lurking in the shadow of the excavator's bucket" (4:143), a nothingness that throttles Arthur before Redrick's eyes. Comparing the first expedition into the Zone that Redrick undertakes (together with Panov) with the last (which, in the company of Arthur, leads to the Golden Ball), one recognizes that the latter adventure has the structure of a "black fairy tale." The fairy-tale quality is not difficult to spot: like a valiant knight-errant seeking the elixir of life or a magic ring, the heroes must overcome dreadful and dangerous obstacles while striving toward a highly valued treasure. Furthermore, Redrick knows that the approach to the Golden Ball is barred by a mysterious "grinder" (4:130), which one must "satiate" by bringing it a human sacrifice. That is why he lets Arthur be the first to approach the sphere—and in fact Arthur dies before his eyes, and his death momentarily breaks the evil spell, so that Redrick in his turn can then reach the Golden Ball. At that point, the authors break off the table and subscribe the word "finis." This, however, is a way out which merely attenuates the shape of things without altering it.

The authors claim—and I have discussed this point with them—that the convergence in the Golden Ball of fairy-tale motif and the Horrific originates solely in the human mind and is a product of chance and human fantasy. Yet, as we have previously stated, one must not arrange all too many "coincidences" that all point in one and the same direction; for it then becomes incredible that they came about by chance. Besides, the last expedition into the Zone does not have the generic attributes of science fiction. The realistic frame for the events transforms itself into that of a fairy tale,[5] because the "coincidences" following one upon the next amount, as we have already observed, to the stereotypic quest for the accursed treasure, though they ought not be identical with any stereotype. The mystery is not consistently preserved to the very end; behind it, the truth keeps shimmering through, since we no doubt have an idea about who the visitors are: they are, once more, monsters, albeit invisible monsters.

The authors attempt to distract the reader from this thought, which flatly forces itself upon us. They stress, for example, that the Golden Ball seen from a distance gives the impression that an unknown giant has accidentally lost it. That, however, is not the correct tactic. It is not the authors' commentary that should divert us from the structurally obtrusive solution, but the events themselves in their objective unfolding. Then, too, the strong impact the epilogue makes spoils the outstanding impression the book makes overall.

Max Frisch transposed the Oedipus myth into our con-

---

5. The degree to which the authors followed the fairy tale's structural pattern in their epilogue can, for example, be seen in the passage in which "black twisted stalactites that looked like fat candles" (4:141) are mentioned. These are all that is left of the people the Golden Ball has killed—that is, all that is left of Redrick's and Arthur's predecessors in the quest for the accursed treasure. In fairy tales such remains—the bones of daredevils who ran out of luck—usually lie at the entrance of the dragon's cave, at the foot of the glass mountain, etc.

temporary reality in his novel *Homo Faber,* wherein the father as unknowingly enters into an incestuous relationship with his daughter as Oedipus did with his mother. Frisch managed the events of the novel in such a way that each possesses a normal, realistic verisimilitude, while together they structurally correspond to the Oedipus myth. The difference between *Homo Faber*'s affinity for myth and *Roadside Picnic*'s for the fairy tale lies herein: that Frisch had in mind the achieved similarity while the Strugatskys by no means desired it. That is the very reason why I say that they "have defeated their own purposes," because only discretion in the arrangement of events could have guarded the end of the story against an unwanted connection with the main plot and hence with the ethos of a fairy tale.

A theologian would have had no difficulty preserving the mystery in *Roadside Picnic,* for he can employ contradictions. But since science does not have such a recourse, it is not an exaggeration for me to say that the difficulties of a fantasy writer who sides with science are generally greater than those of a theologian who acknowledges the perfection of God. . . .

Translated from the German
by Elsa Schieder
and Robert M. Philmus

BIBLIOGRAPHY

Essays

"About Myself," *Poland*, no. 124 (December 1964): 12–13.

"About the Strugatskys' *Roadside Picnic*" ("Posłowie"), Afterword to Arkady and Boris Strugatsky's *Piknik na skraju drogi* (Kraków: Wydawnictwo Literackie, 1977); in English: *Science-Fiction Studies* 10, no. 31 (November 1983): 317–331.

"Cosmology and Science Fiction" ("Science Fiction und Kosmologie"), *Science-Fiction Studies* 4, no. 12 (July 1977): 107–110. Reprinted in *Science-Fiction Studies: Selected Articles on Science Fiction 1976–1977*, ed. by R. D. Mullen and Darko Suvin (Boston: Gregg Press, 1978), pp. 214–217.

"Culture and Futurology" (a chapter from Stanislaw Lem's *Summa Technologiae*), *Polish Perspectives* 16, no. 1 (1973): 30–38.

"A Kind of Credo" ("Eine Art Credo"), *Quarber Merkur* 31 (July 1972); in English: *The Yale Literary Magazine* 150, no. 5 (1984), pp. 1–2.

"Looking Down on Science Fiction: A Novelist's Choice for the World's Worst Writing" ("Science-fiction oder die verunglückte Phantasie"), *Frankfurter Allgemeine Zeitung* (February 22, 1975); in English: *Science-Fiction Studies* 4, no. 12 (July 1977): 126–127.

"Metafantasia: The Possibilities of Science Fiction" ("Zakończenie metafantastyczne"), from Stanislaw Lem's *Fantastyka i futurologia,* tom II (Kraków: Wydawnictwo Literackie, 1970); in English: *Science-Fiction Studies* 8, no. 23 (March 1981): 54–70.

"On the Structural Analysis of Science Fiction" ("Eine strukturalistische SF-Betrachtung"), *Quarber Merkur* 23 (May 1970); in English: as "Introduction to a Structural Analysis of SF," *Science Fiction Commentary,* no. 9 (February 1970): 34–44. Reprinted in *Science-Fiction Studies* 1, no. 1 (Spring 1973): 26–33, and in *Science-Fiction Studies: Selected Articles on Science Fiction 1973–1975,* ed. by R. D. Mullen and Darko Suvin (Boston: Gregg Press, 1976), pp. 1–8.

"Philip K. Dick: A Visionary Among the Charlatans" ("Posłowie"), Afterword to Philip K. Dick's *Ubik* (Kraków: Wydawnictwo Literackie, 1975); in English: *Science-Fiction Studies* 2, no. 5 (March 1975): 54–67. Reprinted in *Science-Fiction Studies: Selected Articles on Science Fiction 1973–1975,* ed. by R. D. Mullen and Darko Suvin (Boston: Gregg Press, 1976), pp. 210–223.

"Planetary Chauvinism: Speculation on the 'Others' " ("Stimmen aus dem All"), *Playboy* (German edition), August 1977; in English: *Second Look* 1, no. 10 (August 1979): 5–9.

"Poland: Science Fiction in the Linguistic Trap" ("Polen: Science Fiction in der linguistischen Falle"), *Quarber Merkur* 20 (August 1969); in English: *The Journal of Om-*

phalistic Epistemology, Supplement No. 1 (August 1969): 1–6. Reprinted in *Science Fiction Commentary,* no. 9 (February 1970): 27–33, and in *Science Fiction Commentary,* no. 19 (January-February-March 1971): 89–94.

"Reflections for 1974" ("Refleksja 1974"), *Kultura,* no. 26 (1974); in English: *Polish Perspectives* 17, no. 10 (October 1974): 3–8.

"Reflections on My Life" ("Mein Leben"), as "Chance and Order," *The New Yorker,* January 30, 1984, pp. 88–98. Reprinted as "Stanislaw Lem, 1921– " in *Contemporary Authors Autobiography Series,* ed. by Cedria Bryfonski (Detroit: Gale Research Company, 1984), I, 255–266.

"Remarks Occasioned by Dr. Plank's Essay 'Quixote's Mills,' " *Science-Fiction Studies* 1, no. 2 (Fall 1973): 78–83.

"Robots in Science Fiction" ("Roboter in der Science Fiction"), *Quarber Merkur* 21 (November 1969); in English: *The Journal of Omphalistic Epistemology,* no. 3 (January 1970): 8–20. Reprinted in *Science Fiction Commentary,* no. 19 (January-March 1971): 117–130, and in *SF: The Other Side of Realism,* ed. by Thomas D. Clareson (Bowling Green, Ohio: The Popular Press, 1971), pp. 307–326.

"Science Fiction: A Hopeless Case—with Exceptions" ("Science Fiction: Ein hoffnungsloser Fall—mit Ausnahmen"), *Quarber Merkur* 29 (January 1972); in English: *Science Fiction Commentary,* nos. 35–37 (July-September 1973): 7–35. Reprinted in *Philip K. Dick: Electric Shepherd,* ed. by Bruce Gillespie (Melbourne: Norstrilia Press, 1975), pp. 69–94. "Appendix: *Ubik* as Science Fiction" reprinted, as "Science and Reality in Philip K. Dick's *Ubik,*" in *A Multitude of Visions,* ed. by Cy Chauvin (Baltimore: T-K Graphics, 1975), pp. 35–39.

"Sex in Science Fiction" ("Sex in Science Fiction"), *Quarber Merkur* 25 (January 1971); in English: *Science Fiction Commentary,* no. 22 (July 1971): 2–10, 40–49.

"The Ten Commandments: Some Remarks on 'Paingod and Other Stories' by Harlan Ellison," untitled letter in *Quarber Merkur* 20 (August 1969); in English: *The Journal of Omphalistic Epistemology*, Supplement No. 1 (August 1969): 6–7. Reprinted, as "The Ten Commandments for Reading the Magazines," in *Science Fiction Commentary*, no. 6 (September 1969): 26; in *Science Fiction Commentary*, no. 19 (January–March 1971): 94–96; and in *Science Fiction Commentary Reprint Edition: First Year 1969, SF Commentary Nos. 1–8*, ed. by Bruce Gillespie (Melbourne: Bruce Gillespie, 1982), p. 106.

"The Time-Travel Story and Related Matters of Science Fiction Structuring" ("Struktura świata i struktura dzieła II: Fantastyka"), from Stanislaw Lem's *Fantastyka i futurologia*, tom I (Kraków: Wydawnictwo Literackie, 1970); in English: *Science-Fiction Studies* 1, no. 3 (Spring 1974): 143–154. Reprinted in *Science-Fiction Studies: Selected Articles on Science Fiction 1973–1975*, ed. by R. D. Mullen and Darko Suvin (Boston: Gregg Press, 1976), pp. 16–27, and in *Science Fiction*, ed. by Mark Rose (Englewood Cliffs, NJ: Prentice-Hall, 1976), pp. 72–88.

"To My Readers," *Poland*, no. 225 (May 1973): 6–9.

"Todorov's Fantastic Theory of Literature" ("Tzvetana Todorova fantastyczna teoria literatury"), *Teksty* 5, no. II (1973); as "Tzvetan Todorovs Theorie des Phantastischen," *Quarber Merkur* 34 (July 1973); in English: *Science-Fiction Studies* 1, no. 4 (Fall 1974): 227–237.

"Unitas Oppositorum: The Prose of Jorge Luis Borges" ("Unitas Oppositorum: Das Prosawerk des J. L. Borges"), *Quarber Merkur* 24 (January 1971); in English: *Science Fiction Commentary*, no. 20 (April 1971): 33–38. Reprinted in *edge*, nos. 5/6 (Autumn/Winter 1973): 99–102.

Interviews

"Amazing Interview: A Conversation with Stanislaw Lem," by L. W. Michaelson, *Amazing Science Fiction Stories* 27, no. 10 (January 1981): 116–119.

"The Future Without a Future: An Interview with Stanislaw Lem," by Zoran Zivković, *Pacific Quarterly* 4, no. 3 (July 1979): 255–259.

"An Interview with Stanislaw Lem," by Anne Brewster, *Science Fiction* 4, no. 1 (n.d.): 6–8.

"An Interview with Stanislaw Lem," by Peter Engel (with John Sigda), *The Missouri Review* 7, no. 2 (1984): 218–237.

"An Interview with Stanislaw Lem," by Raymond Federman, *Science-Fiction Studies* 10, no. 29 (March 1983): 2–14.

"Knowing Is the Hero of My Books," by Andrzej Ziembicki, *Polish Perspectives*, 22, no. 9 (September 1979): 64–69.

"Lem: Science Fiction's Passionate Realist," by Peter Engel, *The New York Times Book Review,* March 20, 1983, pp. 7, 34–35.

"Promethean Fire: An Interview with Stanislaw Lem," *Soviet Literature* 5, no. 239 (1968): 166–170.

"Stanislaw Lem—an Interview," by Daniel Say, *Entropy Negative,* no. 6 (1973): 3–24 (unnumbered). Reprinted, as "An Interview with Stanislaw Lem," in *The Alien Critic* 3, no. 10 (August 1974): 4–14.

"Stanislaw Lem: The Profession of Science Fiction XV: Answers to a Questionnaire," *Foundation,* no. 15 (January 1979): 41–50.

"You Must Pay for Any Progress: An Interview with the Polish SF Writer Stanislaw Lem," by Bożena Janicka, *Science Fiction Commentary,* no. 12 (June 1970): 19–24. Reprinted from *Sovetskaya Kultura,* November 30, 1968.

Letters

June 24, 1970. *Science Fiction Commentary,* no. 14 (August
   1970): 5–6, 20.
January 13, 1972. *Science Fiction Commentary,* no. 26 (April
   1972): 28–29.
"A Letter to Mr. Farmer." *Science Fiction Commentary,* no. 29
   (August 1972): 10–12.
May 7, 1972. *Science Fiction Commentary,* no. 29 (August 1972):
   9, 43–44.
September 9, 1974. *Science Fiction Commentary,* nos. 41/42
   (February 1975): 90–92.
"In Response" (to criticism of his "Todorov's Fantastic Theory
   of Literature"). *Science-Fiction Studies* 2, no. 6 (July
   1975): 16–17.
"In Response (to Professor Benford's remarks on Lem's essay
   "Cosmology and Science Fiction"). *Science-Fiction
   Studies* 5, no. 14 (March 1978): 92–93.
February 6, 1980. *Science Fiction Commentary,* nos. 60/61 (June
   1981): 4.

Reviews

"From Big Bang to Heat Death," extract from "Von Wissenschaft und Pseudowissenschaft," *Quarber Merkur* 52
   (January 1980); in English: *Second Look* 2, no. 2 (Janu-
   ary-February 1980): 38–39. On Paul Davies's *The Runa-
   way Universe.*
"Lost Opportunities" ("M. K. Josephs Roman *The Hole in the
   Zero*"), *Quarber Merkur* 27 (July 1971) and untitled
   review, *Quarber Merkur* 25, (January 1971); in English:
   Science Fiction Commentary, no. 24 (November 1971):
   17–24. On *The Hole in the Zero* by M. K. Joseph, and
   The Left Hand of Darkness by Ursula K. Le Guin.
"Only a Fairy Tale," untitled review in *Quarber Merkur* 31 (July
   1972); in English: *Science Fiction Commentary,* no. 51

(March 1977): 8–9. On Robert Silverberg's *A Time of Changes.*

"On Science, Pseudo-Science, and Some SF" ("Von Wissenschaft und Pseudowissenschaft"), *Quarber Merkur* 52 (January 1980); in English: *Science-Fiction Studies* 7, no. 22 (November 1980): 330–338. On *Other Senses, Other Worlds* by Doris and David Jonas; *Computer Power and Human Reason: From Judgement to Calculation* by Joseph Weizenbaum; *Worlds in Collision* by Immanuel Velikovsky; *Lifetide: The Biology of the Unconscious* by Lyall Watson; *The Martian Inca* by Ian Watson; *The Crash of Seventy-Nine* by Paul Erdman; and *The Third World War: August 1985* by Sir John Hackett.

"Robbers of the Future" (by Sakyo Komatsu), review in *Quarber Merkur* 27 (July 1971); in English: *Science Fiction Commentary,* no. 23 (September 1971): 17–18.

"The Space Flight Revolution" (by William Sims Bainbridge), *Science-Fiction Studies* 6, no. 18 (July 1979): 221–222.

"Two Ends of the World" (by Antoni Słonimski), review in *Quarber Merkur* 57 (July 1982); in English: *The Missouri Review* 7, no. 2 (1984), pp. 238–242.

Books by Stanislaw Lem available in paperback
editions from Harcourt Brace Jovanovich, Publishers